The Power of Standards

Standards often remain unseen, yet they play a fundamental part in the organisation of contemporary capitalism and society at large. What form of power do they epitomise? Why have they become so prominent? Are they set to be as important for the globalisation of services as for manufactured goods? Jean-Christophe Graz draws on international political economy and cognate fields to present strong theoretical arguments, compelling research, and surprising evidence on the role of standards in the global expansion of services, with in-depth studies of their institutional environment and cases including the insurance industry and business process outsourcing in India. The power of standards resembles a form of transnational hybrid authority, in which ambiguity should be seen as a generic attribute, defining not only the status of public and private actors involved in standardisation and regulation but also the scope of issues concerned and the space in which such authority is recognised when complying to standards. This book is also available in Open Access.

Jean-Christophe Graz is Full Professor of international relations at the Institut d'Etudes Politiques, Historiques et Internationales (IEPHI) of Université de Lausanne, Switzerland, and co-founder of the Centre d'Histoire Internationale et d'Etudes Politiques de la Mondialisation (CRHIM). He is also honorary visiting professor at the Department of International Politics at City, University of London.

The Power of Standards

Hybrid Authority and the Globalisation of Services

Jean-Christophe Graz

University of Lausanne

CAMBRIDGE
UNIVERSITY PRESS

CAMBRIDGE
UNIVERSITY PRESS

University Printing House, Cambridge CB2 8BS, United Kingdom

One Liberty Plaza, 20th Floor, New York, NY 10006, USA

477 Williamstown Road, Port Melbourne, VIC 3207, Australia

314–321, 3rd Floor, Plot 3, Splendor Forum, Jasola District Centre, New Delhi – 110025, India

79 Anson Road, #06-04/06, Singapore 079906

Cambridge University Press is part of the University of Cambridge.

It furthers the University's mission by disseminating knowledge in the pursuit of education, learning, and research at the highest international levels of excellence.

www.cambridge.org
Information on this title: www.cambridge.org/9781108499866
DOI: 10.1017/9781108759038

Published with the support of the Swiss National Science Foundation

When citing this work, please include a reference to the DOI 10.1017/9781108759038

First published 2019

Printed and bound in Great Britain by Clays, St Ives plc, Elcograf S.p.A.

A catalogue record for this publication is available from the British Library.

Library of Congress Cataloging-in-Publication Data
Names: Graz, Jean-Christophe, author.
Title: The power of standards : hybrid authority and the globalisation of services / Jean-Christophe Graz.
Description: 1 Edition. | New York : Cambridge University Press, 2019. | Includes bibliographical references and index.
Identifiers: LCCN 2019003493 | ISBN 9781108499866 (hardback) | ISBN 9781108718431 (paperback)
Subjects: LCSH: Service industries. | Service industries–Standards. | Globalization. | Standardization. | BISAC: POLITICAL SCIENCE / International Relations / General.
Classification: LCC HD9980.5 .G724 2019 | DDC 658.5/62–dc23
LC record available at https://lccn.loc.gov/2019003493

ISBN 978-1-108-49986-6 Hardback

Contents

List of Figures — *page* vi
List of Tables — vii
Acknowledgements — viii

1 Introduction — 1

2 The Rise of Transnational Hybrid Authority: A Primer — 24

3 Service Offshoring: The New Frontier of Globalisation — 54

4 Standards as Regulation — 86

5 Doubling Security: Prudential Standards for Insurance Regulation — 114

6 Standards to Create New Insurance Markets — 140

7 The World Office: Standards and Business Process Outsourcing in India — 173

8 Conclusions — 209

References — 223
Index — 246

Figures

3.1 Transnational hybrid authority *page* 82
6.1 Volume of longevity risk transfer deals in the United Kingdom (2007–2017; £bn). 146
7.1 ISO/IEC 30105 and its various parts 205

Tables

3.1 Exports of goods and services, 1990–2017 (US$ at current prices and current exchange rates, in millions) *page* 67
3.2 Share (%) of developing/developed/transition economies of services exports, 1990–2017 (US$ at current prices and current exchange rates) 67
3.3 Estimated world inward FDI flows, by sector and industry, 1990–1992 and 2009–2011 (US$ millions) 68
3.4 Domestic services value added share (%) of gross exports (2011) 69
5.1 Assets by institutional investors in the OECD countries: Ten highest by total amount, 2013, millions of USD 115
6.1 Reporting practices of 34 amongst the biggest insurance and reinsurance multinational companies 166
7.1 Quality and security standards most widely used in Indian ITeS-BPO companies 188

Acknowledgements

This book is the product of several projects that have accompanied me for quite a long time. I wonder whether I shall be able to do justice to all the people and institutions that have contributed to bringing it to fruition. Let us start with the funding body without which nothing would have been possible. The Swiss National Science Foundation (SNSF) awarded me several grants, in particular for a project called Standards and International Relations, which allowed me to build a wonderful team with Christophe Hauert, Eva Hartmann, and Nafi Niang. The SNSF also made it possible for those lines to be read online for free by anyone with access to the Internet, thanks to a commitment to take Open Science seriously. The University of Lausanne, via its programme Vivre Ensemble dans l'Incertain, also gave me a splendid opportunity to gain a sweeping understanding of the arcane world of standard-setting processes by funding the INTERNORM project, designed to support the involvement of civil society organisations in standards-setting procedures related to the International Organization for Standardization (ISO). The project was conceived as an interactive knowledge pool of academic skills and experience gathered by civil society organisations, in particular trade unions, consumer associations, and environmental protection associations. Besides representatives from associations that took part in the project, Christophe Hauert, Danielle Bütchi, Alain Kaufman, and Marc Audétat spared no energy in the steering committee to bring the experiment to fruition. At the University of Lausanne, it is not only state-of-the-art infrastructure and an outstanding campus on the lakeshore facing the Alps which give me a unique working environment; the Faculty of Social and Political Sciences, the Institute of Political, Historical and International Studies (IEPHI), and the Centre of International History and Political Studies of Globalization (CRHIM) provide an inspiring and convivial intellectual environment and have supported me in many ways, including generous research and matching funds to complete this book. As honorary visiting professor at City, University of London, I was also able to spend almost a year expanding on research for the book, writing early drafts, discussing with old colleagues, and meeting new ones. At Cambridge University Press, I would like to thank John Haslam for his support, all the people who helped with the production, as well as the two reviewers whose comments on the book have greatly improved its quality.

In addition to those already mentioned, the following deserve special mention: Jean-Marie Chenou, Claudio Brenni, Nils Moussu, Christelle Genoud, Sylvain Maechler, and Yannick Perticone as teaching assistants and colleagues; Patrick Csikos, Pablo Diaz, Maude Gex, and Pierre-André Blanc as research assistants; Charlie Dannreuther, Patricia Goff, Roger Haydon, Nicole Helmerich, Rahel Kunz, Giovanni Magravitti, Stefano Pagliari, Ronen Palan, and Kees van der Pijl read various parts of the manuscript or earlier versions of the project; Gilles Allaires, Larry Busch, Jean-Pierre Chanteau, Michel Damian, Eve Fouilleux, Oliver Kessler, Anna Leander, Matthieu Leimgruber, Luis Lobo-Guerrero, Allison Loconto, Anastasia Nesvetailova, Andreas Nölke, Pascal Petit, Amin Samman, and Christian du Tertre provided insightful exchanges on numerous aspects closely or loosely related to the power of standards.

Moreover, many invitations gave me the opportunity to test arguments and push further parts and prior versions of the chapters of the book. With no order of preference, and following an inverted chronological order, my warm thanks go to the following: Michèle Rioux, from the Centre d'études sur l'intégration et la mondialisation de l'Univesité du Québec à Montréal; Matthew Paterson, from the Department of Politics at the University of Manchester, and before that for another invitation to the Research Network in International Political Economy of the University of Ottawa; Arnaud van Waeyenberge, from the Centre Perelmann de Philosophie du Droit, Université Libre de Bruxelles (ULB); Jean-Philippe Thérien and Frédéric Mérand, from the Centre d'Etudes et de Recherches Internationales de l'Université de Montréal (CERIUM); Werner Raza, from the Austrian Foundation for Development Research (ÖFSE); Gabriel Siles-Brugge, at the Jean Monnet Centre of Excellence Seminar of the University of Manchester; Auriane Guilbaud and Laurent Jeanpierre, from the Laboratoire des Théories du Politique (LabToP) of the Centre de Recherches Sociologiques et Politiques de Paris (Cresppa); Niilo Kauppi, from the research centre Sociétés, Acteurs, Gouvernement en Europe, Université de Strasbourg; Eve Fouilleux, Allison Loconto, David Demortain, and Pierre-Benoît Joly, from the Laboratoire Interdisciplinaire Sciences Innovations Sociétés (LISIS), Paris; Thomas Lamarche and Christian du Tertre, from the Laboratoire Dynamiques Sociales et Recomposition des Espaces (Ladys), Paris; Marie-Laure Djelic, formerly at École Supérieure des Sciences Économiques et Commerciales (ESSEC), Paris; Klaus-Gerd Giessen, from the

Université Clermont Auvergne; Jacquie Best, from the School of Political Studies at the University of Ottawa; Magdalena Bexell and Catia Gregoratti, from the Department of Political Science, Lund University; Alfred Tovias, from the Davis Institute & Department of International Relations, The Hebrew University of Jerusalem; Marcel Djama, from the Institut du Développement Durable et des Relations Internationales (IDDRI), Paris; Saulo Tarso Rodrigues and Manoel Randolfo, from Federal University of Cuiabá and Unirondon Universit, Brazil; Patrick Le Gallès and Pierre Lascoume, from the Centre d'Études Européennes et de Politique Comparée, Science Po Paris; and Christian Joerges, from the Centre for European Law and Politics, Universität Bremen.

Moreover, I would like to thank everyone who accepted my requests for interviews and facilitated access to various fields of observation. The book would simply not exist without the generosity they showed in the time shared with me. Among all of them, a special thank you goes to Urs Fischer from the Swiss Association for Standardization (SNV), as well as Dana Kissinger-Matry and Reinhard Weissinger from ISO.

Finally, this book also owes much to my friends and family, who endured me and provided support and understanding all through the ups and downs of the long gestation of the project. My mother, Liesl Graz, did not just introduce me to the wider world early in life with my late father; she gave invaluable and timeless help in editing the language of every single chapter. My fantastic wife, Myriam Bickle Graz, was always on my side, even when I was absent for research trips, sabbatical leave, conferences, prolonged writing nights, incomprehensible monologues, and any other unfathomable behaviour. And my three children, Arthur, Zoé, and Gaspard, showed incredible tolerance towards the amount of time that this book has taken out of their life, before seeing it for real and hopefully making some sense of why power may not be where one expects it. For all their love, I dedicate this book to them.

1 Introduction

> *With international trade in services now the driver of economic growth in developed and developing countries, come the dangers inherent in any dramatic market expansion – lack of controls, consumer exploitation, opacity, poor quality, inefficiency, questionable business practices and other obstacles to good service provision. In parallel with such growth, the services sector is in vital need of standards to establish good practice, encourage consistently high service quality, and build consumer confidence.*
>
> Garry Lambert, 'Service with a Smile, Thanks to Standards', *ISOfocus*, #116, 2016:10

According to the quotation at the start of this chapter, taken from the flagship publication of the International Organization for Standards (ISO), there is an upfront rationale for standards supporting the service sector's contribution to growth and development. In the same way as manufacturing is inconceivable without standardised nuts and bolts, it is difficult to imagine providing services across borders without proper guarantees regarding the quality and security of the activity expected to be performed to the customer's satisfaction. From this viewpoint, standards appear to be promising tools against the backdrop of the growing share of services in globalisation. An airline's customer service centre located in the Philippines; legal process outsourcing in India in charge of drafting contracts for law firms in London; reliability of data on natural hazards in Japan, supposedly hedged by reinsurance companies in the United States and Europe; or, more prosaically, requirements for services provided by natural protected area authorities or multinational water utility firms – all these are expected to be specified in mutually intelligible and agreed terms.

A closer look at the importance ascribed to technical specifications in the globalisation of services shows that it reflects a non-conventional form of power in the organisation of contemporary capitalism. Most explanations of the rise of such non-conventional forms of power focus on two interrelated aspects of globalisation: governmental failures in addressing global issues in a world of territorial sovereignty and the ability of large private corporate actors to shape regulatory outcomes and market access in their favour. With a focus on the significance of voluntary international standards as privileged instruments of global governance, this book analyses a third aspect spanning the space between those two poles of public and private power in international relations. Standards set by bodies such as the ISO have long been perceived as narrow technical specifications for organising production, protecting consumers, and facilitating international trade in domains such as measurements, performance, and related effects of manufactured goods. Today, their scope has been extended to non-physical fields such as labour, environment, education, risk and security, or management systems and business models. The opening quotation only gives a glimpse of how great the expectations are for the future. At the same time, standards-setting organisations have mushroomed. While the ISO might not be the best-known organisation of global governance, it fiercely competes with other bodies in a jungle of labels, certifications, benchmarks, and business models.

What non-conventional forms of power do international standards epitomise in the organisation of contemporary capitalism? Why have they become such prominent tools in global governance? Could they be as prominent for the service sector as for manufactured goods? Looking for answers to these questions, a whole body of literature has risen to analyse how market organisation and innovation relies on standards, how standards themselves partake in the diffusion of authority towards private actors, and how this reflects a prevalence of neoliberal ideology in global governance, with all its normative implications for democracy. This book begins from a different perspective, proposing three arguments which can help explain the prominence of such non-conventional forms of power in the organisation of contemporary capitalism: the power of ambiguity, the ambiguity of standards, and the rise of services.

The first proposal is that ambiguity can be seen as a generic attribute of non-conventional forms of power in the regulation of contemporary capitalism. Ambiguity appears as a defining criterion in conferring authority to new actors on a number of new issues across sovereign space in the context of globalisation. We will see how the literature has discussed in great detail the ambiguous status of the private/public divide viewed as a strategic resource for non-state actors to gain power and recognition in global governance. The point here is to suggest that ambiguity not only defines the status of the actors involved in standardisation and regulation but also the scope of issues concerned and the space on which such authority is recognised in complying with standards. This shift in the articulation between the political and the economic spheres across the globe reflects a new topology of global governance from a transnational perspective. This book provides a theoretical and empirical account of this alternative form of authority based on the juxtaposition of instances of power transforming the relation between transnational capitalism and territorial sovereignty – what is considered here as *transnational hybrid authority*.

The second proposal is that the ambiguity of standards accounts for much of their prominence among the various tools of global governance. It allows for highly resilient, multiple, and contradictory policies. It lends itself to ready appropriation likely to support confused lines of accountability. While international standards are often seen as stereotypes that flatten out differences and impose disciplinary power, their role in market organisation and regulation is neither so isomorphic nor inevitably alienating. They can accommodate opposing political economy objectives and power configurations. In theory, nothing would prevent the use of standards by various industries, market actors, and civil society organisations to provide guarantees against opposing understandings of quality and security prospects. In practice, this may rarely be the case, but standards could not be as prominent as they are if they did not convey more ambiguous properties than mere technical specifications set by private firms worldwide. In short, the ambiguity of a transnational hybrid authority goes a long way towards explaining the power of standards and why they have thrived in the organisation of capitalism over the last decade.

The third proposal concerns the global expansion of services. In a context marked by a shift towards a so-called smarter, automated, and more sustainable knowledge-based global economy, services are often defined as the new frontier. They play a key role in supporting integrated production networks and platforms. They are deeply embedded in manufacturing processes depending on all sorts of financial, legal, organisational, marketing, design, or risk management constraints. Accordingly, market access in this domain is less a matter of tariff or investment than of regulation and standards of quality and security requirements likely to have strong social and political implications. Since standards lie at the heart of the service economy, I argue that they also shape the conditions and extent of the convergence likely to support market access. The internationalisation of many types of services has thus become highly controversial in both industrialised and developing countries. Unsurprisingly, regulatory convergence and so-called non-tariff measures lie at the core of negotiations for ambitious and comprehensive preferential trade agreements, such as the Canada–European Union Comprehensive Economic and Trade Agreement (CETA), the aborted Transatlantic Trade and Investment Partnership (TTIP) between the United Sates and the European Union, or the Comprehensive and Progressive Trans-Pacific Partnership (CPTPP) among Asia-Pacific countries. Those negotiations are only three among many examples of how the rise of services defies common expectations on standards and prompts non-conventional forms of power in the regulation of contemporary capitalism.

With a focus on the role of standards in the global expansion of services, this book examines a new form of power in contemporary global political economy. In making sense of the power of standards, its contribution to the existing literature spans five interrelated debates, often at the crossroads of several disciplinary fields.

Globalisation and Transnational Private Authority

The literature on the rise of non-state actors, private authority, and less conventional forms of sovereignty and governance has mushroomed against the backdrop of globalisation. While some globalisation studies continue to oppose states and markets, the approach used here relates to the literature on transnational private authority that views globalisation as a joint process, with new patterns and agents of structural change through formal and informal power and regulatory practices (Cutler et al., 1999; Hall and Biersteker, 2002b; Grande and Pauly, 2005; Djelic and Sahlin-Andersson, 2006; Graz and Nölke, 2008; Krause Hansen and Salskov-Iversen, 2008; Büthe, 2010; Payne and Phillips, 2014; Abbott et al., 2015). Concepts such as power and authority are clearly among the most controversial notions in International Relations, International Political Economy, and cognate fields (Barnett and Duvall, 2005; Guzzini and Neumann, 2012). Moreover, as Lukes (2005) classically pointed out, power is an essentially contested concept, as empirical validation cannot avoid prior normative assumptions. While power and authority are closely related, I do not see them as synonymous. While power needs legitimate social purpose to be exercised by consent rather than coercion, authority conveys an institutionalised form of power that uses formal and informal rules to support such claims of legitimacy based, at least partially, on consent and recognition on the part of the regulated or governed. There is no reason that such mediation should be exclusively associated with government institutions (Hall and Biersteker, 2002a: 4–5). A critical source of non-state authority in globalisation is therefore what Sassen (2003a) calls 'denationalization', i.e. the process which contributes to bringing private and transnational agendas into the political public sphere. As the territorial basis of the state still exists beyond various forms of transnational private authority, relationships between states and non-state actors have become, as Higgott et al. (1999: 6) suggest, 'sometimes conflicting but often symbiotic'. Private authority in international affairs thus presumes at least some consent and state recognition.

Undoubtedly, standards are likely to generate insights into the analytical foundations of such new forms of transnational authority. Yet, with so much emphasis on the actors gaining authority in private regulatory tools, the literature tends to overlook the scope of regulatory practices involved and the reconfiguration of the spatial structure in which such practices are implemented. The nature and the implications of the rise of private actors setting the standards that shape market organisation, access, and regulation across borders calls for an examination not only of who has the authority to set standards but also of what is standardised and where and when standards are implemented, i.e. the actors, the objects, and the space of standardisation. Those three dimensions together shape new forms of power in our societies. They form the backbone of the analytical framework developed in support of my empirical study of standards; they also structure on a more conceptualised level my understanding of the power of standards.

A number of studies use the concept of hybridity to describe the ambiguity implied by such non-conventional forms of authority in contemporary capitalism. According to Hurt and Lipschutz (2016), hybrid rules reflect a new phase of state formation in which state power is enhanced by privatisation and the ensuing depolitisation of the public sphere. In the same volume, Hibou (2016) draws on Weber and Foucault to take the case of ISO standards as hybrid rules supported by neoliberal bureaucratisation. Hybridity takes many forms of attributes of actors and practices involved in – and prompted by – globalisation. Yet, only too often this tells us more about the lack of clearly defined attributes than any distinct features. Still, this is not pointless. Used as a default attribute, hybridity helps to accommodate multiple and contradictory understandings of global governance (Graz, 2008). As Chapter 2 will show, the notion of governance has itself been used in this respect, by enabling the exercise of authority without full control of sovereign rights. Moreover, I draw on insights from semiotics, sociology of science, technology and society, and post-colonial studies to argue that hybridity conveys *substantive attributes* which can help make sense of standards.

With its particular reference to the history of myths, semiotics not only calls to our mind the etymology of the notion; it shows that early representations of collective life used ambiguous meaning and ambivalent values in their power configurations. Ancestral figures of human imaginary were often hybrids; they pervade all sorts of myths' narratives across time and space (Uranie, 1996). Studies in science, technology, and society (STS) put hybrids in the broader context of the crisis of modernity (Beck, 1992). The concept is the cornerstone of Latour's seminal analysis that modernity cannot make sense of what he calls 'quasi-objects' belonging neither to nature nor to society but to both (Latour, 1991). This helps us to understand that standards cannot be confined to the realm of technical specifications and always convey implicit or explicit social values. It is in this regard that STS studies have prompted the so-called practice turn in international relations theory (Best, 2014: 22–25). However, they often lack focus on the proper transnational nature, global reach, let alone the power mechanism of such practices. To some extent, post-colonial studies respond to such shortcomings, not least because debates on hybridity arguably instigated the field itself (Bhabha, 1994; Young, 1995). Their critique of binary relations of power and the emphasis put on subversion and resistance practices at a fluid transnational plane helps to shed light on how standards belong to what Acheraïou (2011: 19) describes as 'syncretic modes of governance'.

In brief, such genuine interdisciplinary thinking allows me to consider the nature and the implications of the rise of private authority across borders in a broader context. I appraise the non-conventional form of power and regulation embodied in standards as a form of authority based on the ambiguous juxtaposition of instances of power transforming the relation between transnational capitalism and territorial sovereignty. In contrast to conventional accounts primarily focused on the rise of new non-state actors in international affairs, the approach used here aggregates three dimensions: the agents defining authority, the issues concerned, and the space of their deployment.

Standards and Regulation

Standards refer here to voluntary technical specifications explicitly documented and published as tools for the organisation of production and exchange of goods and services. Standards codify technical specifications regarding measurement, design, and performances, as well as side effects of products, industrial processes, and services. As seen in the opening paragraph of this book, this includes almost any type of product, process, or service. It can be as down-to-earth as metric and arithmetic definitions, for example the $1/\sqrt{2}$ ratio defined in the ISO international standard used for paper sizes worldwide except in North America (ISO 216). But it also takes in intricate business models qualifying the ability of a firm to disaggregate and complete complex tasks, such as the Capability Maturity Model Integration in the field of services (CMMI for Services) of the CMMI Institute, a technology commercialisation enterprise working as a subsidiary of Carnegie Mellon University and sponsored by the US Department of Defense (see Chapter 7 for more detail). The relatively broad definition of standards given previously acknowledges a difference between formal standards and other norms that arise from unintentional actions and habits (Brunsson et al., 2000; Ponte et al., 2011: 2; Brunsson et al., 2012). Formal standards are set by entities dedicated to such purpose – be they national standard bodies that are members of the ISO, industry-based standards-developing organisations such as those existing in the United States, research centres and management consultancy firms supporting business models, or consortia of firms and organisations working together to develop technical specifications such as the World Wide Web Consortium (W3C) that has designed many web formats and protocols (HTTP, HTML, XML, etc.). Whoever sets the formal standards, expected compliance mechanisms do exist in the form of various conformity assessment processes and certification procedures, with some sort of sanction for non-compliance.

Similarly, the definition used earlier acknowledges a distinction between specifications used in regulations set by public authorities and those that are voluntary and thus formally outside of the authority of the sovereign state. There is, however, considerable overlap between mandatory standards embedded in public regulations and voluntary specifications set by standard-setting bodies. Public authorities have actively encouraged the use of private standards and supported their adoption in mature and emerging technologies (EXPRESS, 2010; National Science and Technology Council, 2011; JISC, 2013; European Commission, 2016e). Moreover, several agreements of the WTO and other trade agreements grant international standards an official status in policies driven towards the harmonisation or mutual recognition of technical specifications used for goods and services (see Chapter 4 for more detail).

Against this backdrop, standards and regulation touch on far-reaching issues beyond mere industrial choices, market failures, technological innovation, and competition, however privileged these are by scholarship in business, economic, and applied-science studies (David, 1985; Vries, 1999; Swann, 2000; Blind, 2004; Swann, 2010; Viardot et al., 2016; Hawkins et al., 2017; Blind et al., 2017). An emerging field of standardisation studies with interdisciplinary backgrounds in history, sociology, organisation studies, law, and political science looks beyond the environment of the firm in order to understand how standards themselves constitute a significant social institution (Krislov, 1997; Brunsson et al., 2000; Tamm Hallström, 2004; Schepel, 2005; Murphy and Yates,

2009; Timmermans and Epstein, 2010; Busch, 2011; Ponte et al., 2011; Yates and Murphy, 2019).

Unsurprisingly, studies in political science put the regulatory power of standards at centre stage. Most of them rely on neo-institutional and comparative political economy approaches analysing the supply and demand of standards and their potential as alternative forms of private and voluntary regulation responding to the transformation of states' traditional role in the economy (Schmidt and Werle, 1998; Mattli and Büthe, 2003, 2011). They have, for instance, provided strong input in the debate opposing the strongly institutionalised ISO and European systems, the more competitive pattern used in the United States, and the oligopolistic nature of so-called consortia standards mostly used in the IT industry (Egan, 2001; Nicolaïdis and Egan, 2001; Tate, 2001). From a political economy perspective, the question is basically that of the relationship between the drive for technical specifications and the institutional framework required to ensure some order in this area at the transnational level. Borrowing concepts such as externalities and transaction costs from public choice and institutional economics, these studies consider to what extent the practices of various agents can be defined by their environments. It follows, so the argument goes, that standardisation provides an institutional guarantee for improving trust in transactions and curbing free-riding risks. For example, Prakash and Potoski have examined the ISO 14000 standards in environmental management systems from a club theory perspective (Potoski and Prakash, 2009; Prakash and Potoski, 2010). ISO standards are thus viewed as excludable, as those not affiliated to the standard cannot benefit from them, but non-rival, as applying the standard does not necessarily diminish the value others gain from applying the same standard. Designing such standards would then always face a trade-off between leniency – increasing their acceptability and ultimately the number of firms in the club – and stringency – insuring the credibility of the club to produce externalities on its own. Such accounts indisputably help to formalise determining factors of cooperation and conflict underpinning the institutional framework of standards used to differentiate markets. Their limitation, however, rests on a managerial approach focused on firms' and broader stakeholders' utility maximisation functions. In other words, I agree that standards are resources to differentiate markets, but this differentiation is not only the result of a utilitarian rationality calculus implemented by firms and stakeholders. By implying that the logic of action trumps its content, the understanding of the power relations involved in standardisation is thus confined to quantifiable and *a priori* defined criteria based on utilitarist assumptions.

In contrast to a narrow focus on institutional environments and the logic of collective action, I opt for an analytical strategy that stresses the ambiguous content of power relations in the regulatory authority of standards, their evolving variety across borders, and the ongoing struggles to set and conform to them. The non-conventional form of power and regulation established by standardisation brings to mind what Cox called the internationalisation of the state appropriated to the internationalised process of economic policy harmonisation and, more specifically, the *nébuleuse* of official and unofficial networks, with representatives of business, the state, and academia working towards the formulation of a consensual policy for global capitalism (Cox, 1987: 262, 2002: 33). From the broadest sociological point of view, as we saw earlier, standards are a social institution supporting a distinct form of domination. As shown by scholars from the French regulation school, they also call up the institutional economics of John Commons: in contrast to price signals or intrinsic attributes of goods or services, standards result from power relations and are here to qualify objects in such a way as to eventually control the individual action of agents involved in economic transactions (Commons, 1934; Chanteau, 2011; Allaire and Lemeilleur, 2014). Likewise, Timmermans and Epstein (2010: 83) observe that, 'somewhere between glorified globalization and dark dehumanization, each standard achieves some small or large transformation of an existing social order'. For his part, Busch (2011: 2) emphasises that 'standards shape not only the physical world around us but our social lives and even our very selves. [... They] are recipes by which we create realities'. From a legal perspective, Schepel (2005: 4) reminds us that the public or private nature of standards and the space of their deployment overcome conventional oppositions: 'Standards hover between state and the market; standards largely collapse the distinction between legal and social norms; standards are very rarely either wholly public or wholly private, and can be both intensely local and irreducibly global. ... standards can be seen as links between these spheres and institutions'. In the same vein, a great deal of scholarship on the rise of private authority in political science and global political economy no longer sees standards as outright privatisation and deregulation. Instead, the phenomenon is perceived as part of the broader organisation of the capitalist system (Murphy and Yates, 2009), or a 're-articulation of governance', in which public regulation has 'retreated in some areas of the economy, but at the same time other forms of governmental and inter-governmental regulation are actually being strengthened' (Ponte et al., 2011: 7). As Hauert (2014: 2 – my translation) emphasises, 'the influence of those private arrangements in various institutional environments, their relationship with public authorities and the characteristics of actors supporting it remain largely ambivalent'. This is even truer with regard to services, for which profound cultural tenets, societal values, and labour issues are hard to pin down (Allen and du Gay, 1994). Moreover, with the advancement of deregulation, liberalisation, and privatisation, new service standards are likely to compete with previous rules governing public utilities, and more generally the social foundations of state power.

My analysis of transnational hybrid authority draws from such insights to set out a three-dimensional framework for a critical and comprehensive picture of the range of actors involved in setting standards, the breadth of the issues concerned, and the extent of deterritorialisation of standards recognition. Together, these overcome conventional understanding that opposes mandatory regulation and voluntary standards, technical specification, and social values and institutions, as well as the territorial space of the sovereign state and the borderless world of global markets. Moreover, in contrast to studies which oppose the profoundly institutionalised European and ISO environment to the weaker and highly privatised US system, I argue that competing models of standardisation do not reproduce such territorial and institutional determination. Instead, they reflect contrasting types of relationships between standards and society at large. International standards – as hybrids – are ambiguous and double-edged. They can be used either as driving forces to broaden the domain of market self-regulation, or as alternative instruments for embedding markets within society. Accordingly, the institutional developments of service standards are likely to require trade-offs between advocates of the commodification of technical standards across borders and promoters of further socialisation of international standards as applied to distinct and well-chosen service sectors.

Globalisation and the Rise of Services

In addition to furthering understanding of the peculiar power of standards in the rise of transnational private regulation, this book provides an innovative account of the relationship between globalisation and the rise of services, with a focus on the neglected role that standards play in this regard. An economy based on information and knowledge not only increases the share of services, but also the tradability of activities previously viewed as requiring a face-to-face environment. Besides the globalisation of traditional service activities such as tourism, transport, banking, and insurance, global production and market networks become increasingly reliant on service offshoring – i.e. the purchase of services abroad or the transfer of particular tasks to a foreign location that makes the most of the management mantra 'your mess for less'. Service offshoring as such remains a highly debated issue, especially concerning the balance between alleged economic benefits and political costs following significant job losses in rich countries. In fact, the internationalisation of services has often turned out to be less momentous than expected and limited to specific sectors and key emerging economies such as India.

Conventional explanations of the internationalisation of services focus on a number of drivers and barriers of trade in services, among which ICT and labour costs arbitrage come first (see for instance: Bryson and Daniels, 2007b; McIvor, 2010; Dicken, 2015). Other drivers of the internationalisation of services include the growing mobility of people, with more consumers buying services abroad (e.g. so-called old-style tourism plus the fast growing market of health tourism) and more service providers posting workers abroad (as provided for by the so-called mode 4 of the GATS on the movement of natural persons, the EU Directive on posted workers, the H-1B visa in the United States, and a number of other pieces of national legislation). Other determining factors of the internationalisation of services include language, cognitive, and cultural skills used across industries.

Many analyses, however, emphasise that the idiosyncratic nature of services can make them important obstacles to trade. The production and exchange of services do indeed differ from that of manufactured goods. The notorious difficulty of defining services will be dealt with later (see Chapter 3 for more detail); suffice it here to note that the trouble involved in having a shared representation of expected quality goes some way towards explaining the difficulty of internationalising the market of many sorts of services, let alone establishing a market in the first place. The economic literature on asymmetries of information and market uncertainties has looked at how this has far-reaching consequences for services and their internationalisation (Akerlof, 1970; Grönroos, 1990; Zeithaml et al., 1990). While seeking to identify measures likely to be trade restrictive or market destructive, it moreover recognises that the resulting pervasiveness of services regulation includes standards (Copeland and Mattoo, 2008; World Trade Organization, 2012). It gives little attention, however, to the role of standards in the internationalisation of services beyond market failures, market access, and intergovernmental cooperation in trade policies. As seen earlier, employment, trade, and investment in the domain of services are often less a matter of price mechanisms, tariffs, and investment than regulation and standards to be assessed against distinct quality performances, security guarantees, and protection of consumers, likely to have strong and conflicting social and political implications.

The few studies specifically focused on the relation among the rise of services, their internationalisation, and standardisation usually consider that the ability to set services standards supporting internationalisation is a function of key sectorial and institutional specificities. Many microeconomic studies examine which services are likely to be standardised according to taxonomies determining firms' choices between standardisation and customisation (Boden and Miles, 2000b; Blind, 2004; Djellal and Gallouj, 2010). Scholarship inspired by the French *régulation* theory and economic sociology has a broader understanding of the social and political issues at stake in the standardisation and internationalisation of services (Callon et al., 2002; Du Tertre, 2002; Gadrey, 2003; Petit, 2007; Du Tertre, 2013). In considering that standardisation and internationalisation are closely defined by the attributes of labour relations, forms of competition, and domestic institutions affecting services, existing studies suggest a restrictive hypothesis that paradoxically tends to conceal a number of political economy power plays.

In contrast to analyses based on sectorial and institutional specificities, my approach emphasises an extensive hypothesis. By linking the global marketplace to distinct national economies, service standards can respond in various ways to quality and security uncertainties. The prospects of greater market integration for services then depend on such non-state arrangements. As mentioned previously, their power configuration can accommodate opposing political economy objectives. Standards may in many cases be stereotypes that deny cultural and labour issues involved in the service economy. Yet, they can also contribute to a more progressive understanding. For example, they can give guidance on occupational health and safety as in ISO 45001 or provide highly specific requirements on the construction of scaffolding and thus help avoid unnecessary hazards for workers on building sites (Bergström, 2004). The concept of transnational hybrid authority suggests that standards *per se* neither support nor hinder the internationalisation of services. It all depends on which actor exerts authority to set standards on what issue and across which jurisdictions they gain recognition. According to my three-dimensional analytical framework, this involves actors with opposing political economy interests and values, concerns issues which intrinsically blur the frontier between societal and physical worlds, and relies on a system of certification and accreditation that reinforces the deterritorialisation of sovereignty. Basically, the ambiguity underpinning the authority of standards lays the very basis for the study of their role in the internationalisation of services.

This book further examines the links between the power of standards and the rise of transnational private authority, competing models of regulation, and the worldwide expansion of services, with in-depth studies of two contrasted service sectors and activities. It thus also contributes to the existing literature on each case.

Insurance and the Financialisation of Contemporary Capitalism

The book considers insurance services as key market integrators closely related to the financialisation of contemporary capitalism. Insurance often appears to be tedious; in reality, it is an unwitting giant of global finance and a key institution of informal governance and alternative sovereignty. Together with pension funds, to which the life insurance industry is closely related, insurance companies account for more than half of total institutional assets under management in OECD countries. If we look at the total amount of money spent in insurance premiums in 2015, we come across the extraordinary figure of around 8 per cent of GDP in advanced markets, with some $ 3,500 of premiums paid per capita – more than four times as much as worldwide defence spending as a percentage of GDP (International Institute for Strategic Studies, 2016: 490; Swiss Re, 2018b: 37). Beyond figures alone, insurance services are closely related to the post-crisis accumulation regime of financialised capitalism. With shrinking long-term lending by banks and austerity policies set to be around for some time, the insurance industry plays a key role in financing long-term investments such as in infrastructure, innovation, education, and health. Moreover, with pension schemes and pension funds using ever more complex financial products offered by life insurers and investment banks, insurance companies are slowly but surely gaining centre stage in the choices that our societies face with the challenge of an ageing population.

Against this backdrop, the extent to which the significance of this industry has been neglected outside the field of actuarial studies is quite remarkable. Studies in history and sociology have examined a number of concrete practices in various insurance lines. They apply the concept of governance to appraise discursive regimes and governmental rationalities of moral and societal risks either from a Foucauldian perspective or to investigate further Ulrich Beck's hypotheses on the emergence of the risk society (Beck, 1992). They typically focus on the domestic realm and more particularly on the United States, Canada, or the City of London (see in particular: Baker and Simon, 2002; Ericson et al., 2003; Aradau et al., 2008; Collier, 2008; Lengwiler, 2009; Clark et al., 2010; Doyle,

2011). Very few studies have investigated the pioneering hypotheses of the late Susan Strange and Virginia Haufler on the ambiguous authority of the public/private nexus of insurance services across domestic and global realms (Strange, 1996: 122–134; Haufler, 1997). The case of insurance in climate change policy (Paterson, 2001; Haufler, 2009; Grove, 2010) and post-structuralist studies on life insurance (Lehtonen, 2014; Lehtonen and Van Hoyweghen, 2014) remain exceptions which prove the rule. Also of note is Zhang's (2014) insider's account that provides a non-mathematical yet critical analysis of life insurance regulation, principally in the United States. Last but not least, Lobo-Guerrero's (2011, 2012, 2016) inspiring trilogy combining Foucauldian approaches, security studies, and international political economy provides a momentous contribution to our understanding of the various ways in which insurance can be a global and powerful technology of government to create an infinite space for market development, to promote and protect distinct lifestyles – that is, essentially to create an alternative form of sovereignty.

Those few accounts help us to understand the extent to which private insurance contracts rely on a flurry of institutional designs and norms of behaviour in their provision of security on a scale that transcends states' territorial sovereignty. The emphasis given in this book is, however, more specifically focused on the standards sought by the insurance industry in its ability to control, transfer, and distribute risks across borders, let alone to steer clear of state intervention as far as possible. With a distinct focus on standardisation processes, the book carries on its aim to probe the ambiguous authority of standards likely to support the expansion of the tertiary sector. Conventional accounts would identify the insurance industry as a most-likely case of services internationalisation and standardisation, as it includes financial activities that are far from the ideal type of highly relational and intangible services. In contrast with such accounts, my in-depth case studies show that setting standards for the insurance industry remains difficult and contentious, even as they have become key instruments of market regulation and creation. The book analyses in detail why standards are in a better position to serve as regulation in the post-crisis-era, in particular in the wake of the European Directive Solvency II, considered the most ambitious regulatory overhaul ever undertaken for insurance industries (Chapter 5). It also shows how the insurance industry relies on standards to create new markets by making new objects insurable to an ever-larger part of the world's population, with particular focus on the securitisation of life insurance, reinsurance of natural catastrophes, and formats of data exchange and non-financial reporting (Chapter 6).

Business Services, Development Policies, and India

India's achievement as the top business services location in the world provides a vivid counter-intuitive case study of the role played by standards in the internationalisation of services. Many activities that have made India the world office are close to the ideal type of highly relational and intangible services, precisely those that conventional accounts take as less likely to be standardised and internationalised. While the industry has its roots in the repetitive tasks of software coding, basic back-office tasks such as data processing, and call centres for customer relations, it now includes cutting-edge business services with a highly skilled workforce in a wide range of activities such as legal, fiscal, banking, insurance, medical, architecture, and consulting services. These delocalised activities are generally called IT-enabled services (ITeS); when the specifics of the tasks outsourced are seen as particularly important, the terminology commonly used is business process outsourcing (BPO).

Studies painting a rosy picture of the success story of the Indian service industry worldwide have mushroomed over the last decade or so. On the positive side, Friedman's landmark account portrayed India as the exemplary case of the new 'flat world' of the globalised twenty-first century, in which entrepreneurial spirit matches proper use of information and communication technology and a skilled workforce to create 'the possibility of a new form of collaboration and horizontal value creation: outsourcing' (Friedman, 2006: 131). A darker side was shown by studies emphasising that the industry depends on deeply exploitative labour relations that look like assembly lines 'in the head' (Taylor and Bain, 1999) and are part of a broader neoliberal development regime (Upadhya, 2009). Moreover, most analyses remain stuck in a narrow state–market divide when discussing India's development policies supporting the IT sector and business services. While market-driven accounts consider the liberalisation policies adopted in the early 1990s as a turning point (Heeks, 1996; Nayyar, 2012: 48ff), opposing views focus on the role of the developmental state in technological innovation for late industrialised economies. In contrast to those narratives, the focus on standards lays emphasis on a wider range of market institutions that have helped India to become the world's office. A few studies use an evolutionary political economy perspective to emphasise the processual, sequential, and overlapping dimensions of the wide range of institutions that shape technological trajectories such as the development of the Indian service industry (Dossani and Kenney, 2007; Parthasarathy, 2013b). None, however, specifically focus on the significance of technical standards as key instruments for either reinforcing or overcoming path-dependent advances of the industry.

The story told in Chapter 7 shows that standards have mattered from the very start of the journey to the latest prospects of the industry. It was the lack of standards in the nascent Indian IT hardware industry that prompted the unexpected emergence of IT services. Several decades later, the publication in 2016 of the ISO/IEC 30105 standard 'Information technology – IT Enabled Services/Business Process Outsourcing (ITESBPO) Lifecycle Processes' demonstrates the ability of Nasscom, the voice of the IT service industry in India, to initiate the adoption of a new standard specifically dedicated to the BPO industry and thus, for the first time, to overcome path-dependencies and make the Indian business service industry a standard maker rather than mere standard taker. For scholars who would see standardisation in such ideal-typical intangible and relational service industries unlikely, this is clearly counter-intuitive and supports my extensive hypothesis on the power of standards in the globalisation of services. Moreover, in contrast to the existing literature on business models and quality standards, my analytical framework shows that the power of those standards is more ambiguous than usually assumed in terms of public involvement, societal implications, and territorial recognition.

Methods

This book not only provides a framework for a critical analysis of the transnational hybrid authority of standards and their role in the rise of services in the globalisation of contemporary capitalism. It also offers a detailed empirical study of the institutional environment of standardisation and in-depth enquiries on two contrasted service sectors and activities. Considering the breadth and diversity of the service economy, a controlled contrast between cases is necessary to assess the extent to which the proposals and hypotheses made in this book are relevant. I applied a method of maximum variation purposeful case sampling with a view to combining cross-institutional and sectoral analyses (Patton, 2002: 230ff). In qualitative methods, purposeful sampling is a privileged means of identifying information-rich cases in order to single out common patterns of particular interest with regard to the hypotheses guiding the research.

The research targets the most important institutions involved in the authority conferred on standards. Particular attention is given to initiatives that take place in the ISO environment and its relations with the World Trade Organization, the European standardisation system, and the distinctive mechanisms that exist in the United States. Yet, as the range of organisations involved in setting and assessing conformity to standards is much wider, the research includes other standardisation bodies, research centres, and management consultancy firms supporting business models, as well as consortia of firms and organisations working together to develop technical specifications such as those mentioned earlier.

As far as the sectorial basis is concerned, I selected my cases so as to cover the extreme heterogeneity of activities identified as services. To identify what stands out in current and future standards developments across widely diverse forms of services, my sampling targeted either high or low values on the main characteristics differentiating the service economy. In order to reconcile conventional distinctions based on categories such as business/non-business services or services to households/companies with critical approaches focused on productive configurations between labour, technologies, and organisational mechanisms, I used the four following criteria:

1. Relational intensity: transactions in services, in contrast to goods, imply an effect of the recipient on the provider's behaviour; however, depending on the sector and the organisational structure chosen to provide the service, the intensity of the relation between the recipient and the provider may vary a great deal, ranging from professional counselling to transport logistics.
2. Immateriality: the types of 'support' targeted by the action of the service differ considerably; they can range from the very material (e.g. objects to be maintained or financial assets valorised) to largely immaterial (e.g. individuals to be counselled, coded information processed, or organisations managed).
3. Consumers' implication: despite an ever-increasing complexity of productive configurations involving all sorts of intermediaries and outsourcing processes, services can still be distinguished between business services, whose transactions target the business community, and consumer services, directly implying the consumer as an end-user.
4. Labour intensity: in a context of massive industrialisation in the service economy driven by information and communication technologies, services can involve large amounts of capital (e.g. transport) but still mostly rely on skilled or unskilled labour (as in consulting or call centres).

I built upon such criteria a maximum variation matrix of potential cases that I used to identify a homogeneous sector-wide basis on which to probe my hypotheses. Among numerous sectors and sub-sectors with varying scores from criteria to criteria, two stood out as displaying a relatively straight alignment of either high or low value scores on all four characteristics: call and customer centres exemplify areas with high relational intensity, immateriality, end-user-orientation, and labour intensity; in contrast, insurance epitomises activities with low relational intensity, greater materiality, a strong business-oriented implication, and capital intensity. While the choice to focus on those two industries results from a controlled contrast sampling strategy, early desk and field research on customer centres promptly led me to realise that the industry had come a long way since its early start as cheap call centres located in remote locations such as India or Morocco. As seen earlier, there are no bounds for outsourcing highly complex and disaggregated tasks. Just as early customer centres have made room for IT-enabled business process outsourcing, industry characteristics have become less oriented towards end-users and less labour intensive – thus loosening the strictness of the initial sampling scores. Yet, relational intensity and the immateriality of the support targeted by business process outsourcing (i.e. the capability and the maturity of an organisation to complete distinct tasks) are so great that they still differentiate in a well-contrasted way the sector as compared to the insurance industry.

The data collection relies on documentary sources, large data sets, field research, participatory observation, and interviews with industry experts and high-ranking officials over many years. While I have been studying standardisation since the early 2000s, extended field research on service standards was undertaken from 2006 to 2017 (either on my own, together with PhD students and post-doc researchers, or by them alone). More than 200 semi-directive interviews provided insights from key players in the standardisation bodies, companies implementing them, civil society organisations, regulatory agencies, and government ministries in relevant sectors in Switzerland, France, Germany, the United Kingdom, the Netherlands, Sweden, the European Union, the United States, Morocco, and India.[1] Moreover, an in-depth

[1] Interviews include high-ranking officials of all major bodies of the international, American, and European context of standardisation: the American National Standard Institute (ANSI); the American Society of Mechanical Engineers (ASME); the American Society for Testing and Materials (ASTM international); the National Institute of Standards and Technology (NIST); the Consumer Electronics Association (CEA); the Consumer Specialty Products Association (CPSA); the International Organization for Standardization (ISO); the Comité Européen de Normalisation (CEN); the Association

understanding of the intricate world of standardisation was considerably reinforced by the lessons drawn from a pilot project designed as action-research based on a platform pooling academic skills and civil society organisations to strengthen their participation in standards setting (INTERNORM, funded by the University of Lausanne, Switzerland, from 2010 to 2014). The project gave the research team direct access to 11 ISO expert groups in both nanotechnologies and tourism services; after more than a total of 45 days spent in ISO technical committees and more than 150 comments and drafting recommendations submitted in this context, INTERNORM gave me a unique understanding of how standard setting bodies work in concrete terms (Graz and Hauert, 2019).[2]

The Book in Brief

Following this overview, Chapter 2 engages with theories of global governance and private regulation to explain how and why standards support what I call a transnational hybrid authority. To explain this, I set out to respond to the following three questions. First, why is reference to the notion of hybridity ubiquitous in describing the power of standards as a distinct form of global governance? I argue that references to hybridity in contemporary debates on globalisation, regulation, and governance offer a handy *default attribute* likely to accommodate multiple political, ideological, and technical exigencies. A brief overview of recent scholarship on regulation and a more detailed genealogy of the concept of governance show the extent to which this reflects the importance of ambiguity in governance policies to exercise power without the plain attributes of sovereign rights. The second question is what, then, is the power of standards. I argue that, beyond a mere default attribute, the concept of hybridity gives credit to ambiguity that should be understood as a *substantive attribute* of the non-conventional forms of power and regulation embodied by standards in the context of globalisation. This non-conventional form of power and regulation is conceived as a *transnational hybrid authority* and defined as a form of authority based on the ambiguous juxtaposition of instances of power transforming the relation between transnational capitalism and territorial sovereignty. The argument is based on discussion and critique of literature in semiotics, sociology of

Française de Normalisation (AFNOR); the British Standards Institution(BSI); and the Deutsches Institut für Normun (DIN).

2 For more information: www.unil.ch/vei/internorm.

science and technology (STS), and post-colonial studies. The third question is how all this plays out, i.e. how is this non conventional form of power exerted by setting and conforming to standards on multifarious old and new issues across different sovereign spaces? While most scholarship in international relations and international political economy is focused on the rise of private actors, I show that ambiguity confers authority not only on new actors but also new issues across sovereign spaces. The final section thus outlines the analytical framework including the three dimensions of actors setting standards, the scope of the standards, and the space on which such authority is recognised. This three-dimensional analytical framework will be used throughout the book to study the power of standards and its implications for broader issues shaping and transforming the relationship between transnational capitalism and territorial sovereignty.

Chapter 3 examines the relationship among globalisation, the expansion of the tertiary sector, and the growing authority conferred on standards in order to situate opposing arguments on the potential role of standards in supporting the globalisation of services. It first provides a contextual and conceptual background on services, the knowledge economy, and the service/manufacturing overlap in integrated production and market networks based on relevant literature in political economy. In discussing the so-called 75/25 puzzle, this chapter fleshes out why the overall share of services in the global economy has not considerably changed and continues to represent around 25 per cent of world trade, despite the fact that services account for 75 per cent of GDP and employment in rich countries. While restrictive hypotheses on the internationalisation of services in institutional economics and regulation theory focus on the informational, institutional, and sectorial factors hindering trade transactions in this domain, I contend that a perspective inspired by evolutionary and international political economy allows for a more extensive hypothesis that sheds light on the potentially greater importance of standards. As service offshoring is less a matter of tariffs than of regulation and standards, I next describe more precisely how international standards reflect a form of transnational hybrid authority that defines a wide range of quality and security requirements likely to have strong social and political implications.

While Chapter 2 sets the theoretical framework and Chapter 3 provides background and further analytical insights on the relation between globalisation, services, and standards, Chapter 4 is a journey around the various institutions providing authority to standards as *de jure* or *de facto* regulatory instruments governing the internationalisation of services. It analyses how the regulatory framework of law has yielded ground to voluntary standards drafted by a raft of international or regional public and private sector bodies. The reader will start the trip in Geneva, with some background on the General Agreement on Trade in Services (GATS), other service-related WTO provisions, and the International Organization for Standardization (ISO). After those privileged arenas for the development of service standards, we will move to Brussels to present developments on service standards in the European context, from the so-called 1985 New Approach to the new EU Regulation on Standardisation entered into force in 2013 and the ensuing standardisation package announced in 2016. Next, we will cross the Atlantic to examine to what extent the United States is a special case of standardisation as it is usually understood. While the European system of standardisation tends to rely on territorially based legitimacy and state oversight, the American system gives preference to competing sources of standards and relies on market mechanisms. The analysis re-examines this conventional view of a transatlantic divide in standardisation: it shows that contrasting models of standardisation do exist, not only between but also across those systems, and that the variance between product and service standards is much greater in the European context and the ISO system than in the United States, where it is hardly debated. The chapter next crosses oceans again, with some forward-looking discussion of the crucial role played by standards in regulatory convergence and non-tariff measures in the new generation of preferential trade agreements. The chapter finally recaps the argument regarding the institutional ambivalences of service standards along the three core dimensions of the agents involved, the issues concerned, and the space in which such standardisation processes are likely to be recognised.

The three following chapters provide in-depth studies on two contrasted service sectors and activities. Chapter 5 is the first of two chapters focused on standards likely to support the internationalisation services in what conventional accounts identify as a most-likely case (the case of the insurance industry being far from the ideal type of relational, non-material services). In contrast to such accounts, both chapters show that setting standards for the insurance industry remains difficult and contentious, even if they have become key instruments of market creation and regulation. This first chapter is focused on the regulation side of the insurance industry in the post-crisis era. It first provides some background on insurance services, their close relation to the financialisation of contemporary capitalism, and why they have become a significant institution of informal governance and alternative sovereignty. It then explains why standards are in a position to stand as regulation in the shift towards a risk-based regulation that has taken place over the last two decades and that has not been opposed in the post-crisis context. I then examine in more detail the astonishing power that the European Directive Solvency II reflects in this regard. This most ambitious regulatory overhaul ever undertaken for insurance industries has tremendous implications across the industry and way beyond the European Union. I show in particular that the change in the game made by Solvency II is as much about the power of the regulators as it is about conferring authority on standards and internal models. Subsequently, I show how Solvency II set the stage for developments at the global level under the aegis of the International Association of Insurance Supervision (IAIS) and regulatory policy reforms in the United States.

Chapter 6 continues by looking at insurance standards used in market creation rather than those associated with market regulation. It touches on the extent to which the insurance industry relies on standards to create new markets by making insurable new objects to an ever-larger part of the world's population. To this end, the chapter unveils a number of little-known standards that are nevertheless indispensable to the functioning of insurance markets. My enquiry focuses on how standards are instrumental in pushing the frontier of highly innovative and securitised insurance markets ever further, with a distinct focus on life insurance and its close connection to pension policy reforms in the post-crisis environment. I also examine how heavily existing insurance markets rely on standardised formats of data exchange and non-financial reporting, with special focus on the difficulties of reinsurers in establishing common standards to hedge the accumulation of risks associated with natural catastrophes.

Chapter 7 turns to an opposite case study of the internationalisation and standardisation of services by looking at a least-likely case of highly immaterial and deeply relational services. It provides an in-depth analysis of India's achievement as the top business service offshoring location in the world and of the significant role played by standards. It starts with some historical background on how India became the world's office, emphasising how standards played a crucial role in the emergence of a wider spectrum of market institutions than those usually accounted for by the state–market divide of the existing literature. It then examines in more detail the rise and range of international standards and certified management tools used in business process outsourcing in India. In contrast to conventional accounts that relational and intangible services are hard to standardise and, hence, internationalise, the analysis sheds light on the prominence of service standards in India and their ambiguous authority. Finally, the chapter focuses on the particular role of Nasscom, the voice of the Indian IT service industry, from the time when service offshoring began scaling up, turning it from a mere standard taker to a world-class standard. A particular case in point is the successful sponsorship of a new ISO standard for business process outsourcing services.

The conclusion recalls the basic assumption that informs this book and its key arguments and findings, before drawing broader implications on the power of standards.

2 The Rise of Transnational Hybrid Authority

A Primer

For the 30,000 or so experts who participate every year in the technical committees of the International Organization for Standardization (ISO), standards may look like well-tuned instruments for defining requirements, specifications, guidelines, or characteristics used in the production, exchange, and consumption of goods and services, and more generally in the functioning of organisations. Yet, for scholars studying recent transformations in the global economy and non-conventional forms of power and regulation in contemporary capitalism, standards often look like UFOs, even as their significance is increasingly recognised. Scholars from many quarters of social sciences have increasingly used the word 'hybrid' to unpack such a complex blend. For instance, in a comprehensive account of the regulatory strategies and institutional arrangements adopted by the European Union in promotion of the Single Market, Egan associates the role of standards to a 'distinctive model of regulation that is a hybrid of state and non-state actors' (Egan, 2001: 264). A decade later, Bartley recognised that a great deal of work remained to be done to fully understand concepts such as the 'complementarity, rivalry, and hybridity in the interplay of multiple standards' (Bartley, 2011: 519). Why does the term 'hybrid' continue to crop up in reference to the growing power of standards in the disparate phenomena closely or loosely related to globalisation and global governance? What are the common attributes of all these phenomena? Could 'hybrid' be something more than a general attribute? If so, what are the substantive attributes of what I will shortly refer to as transnational hybrid authority? Finally, how does all this work in practice – that is: who standardises what and where?

Through those *why*, *what*, and *how* questions, this chapter discusses the non-conventional forms of power and regulation enacted by standards in international relations and the global political economy. The chapter makes three broad arguments. First, in answer to *why* the term 'hybrid' is so frequently used to describe the power of standards and their relations to new forms of global governance, I suggest that references to hybridity in contemporary debates on globalisation, regulation, and governance are often made to justify the highly volatile, and sometimes contradictory, policies that are required to accommodate multiple political, ideological, and technical exigencies. Just as the concept of governance allows for the exercise of authority without the full attributes of sovereign power, the notion of hybridity gives credit to the ambiguous sources of legitimacy that global governance draws on in order to wield authority beyond sovereign control. A brief overview of recent scholarship and a more detailed genealogy of the concept of governance will show how the notion of hybridity has been predominantly used as an unspecified general attribute – a 'default attribute' – that leaves such non-conventional forms of power virtually undefined.

The second argument put forward in this chapter answers the *what* question – what are the substantive attributes of the power of standards? It makes the case for understanding the concept of hybridity as a 'substantive attribute' of ambiguity; in other words, as an ontological property shared by a majority of the non-conventional forms of power that have arisen in conjunction with contemporary global capitalism. Just as a piece of collage artwork acquires a unique aesthetic status from a paste-up of assorted materials – think of fragments assembled by Picasso, Duchamp, Schwitters, and their like – so do hybrid phenomena acquire an identity of their own. To pursue the analogy further, our *pieces rapportées* are supplied by different areas of enquiry, particularly in semiotics, the sociology of science, technology, and society, as well as in post-colonial studies. The non-conventional form of power and regulation exercised by standards is conceived as a *transnational hybrid authority* and defined as a form of authority based on the ambiguous juxtaposition of instances of power transforming the relation between transnational capitalism and territorial sovereignty.

Moving on to the *how* question, I look to how a host of new actors rely on this notion of ambiguity to exert non-conventional forms of power in standardising multifarious old and new issues across different sovereign spaces. In political science, international relations (IR), and international political economy (IPE), this ambiguity is identified with the blurring of the private/public divide, and the literature on private regulation offers extended accounts on how this blurring is strategically employed by non-state actors to gain power and recognition in global governance. What is missing, however, is the recognition that ambiguity imbues not only the status of the actors involved in standardisation and regulation but also the scope of the issues on which they operate and the spaces on which they exert their authority. All three categories, i.e. subjects, objects, and spaces of authority, will be charted in a three-dimensional framework to analyse transnational hybrid authorities – or 'hybrids' for short. Such a framework will drive my study of the ambiguous and multi-layered instances of power conveyed by international standards. It could also hold for understanding broader issues shaping and transforming the relationship between transnational capitalism and territorial sovereignty.

Why Hybrids Now?

Why are standards so often defined as an archetypical example of hybrid regulation? Does this refer only to the technical complexity involved in the all-pervasive use of such market and regulatory instruments whilst seeking global solutions to global problems? Is it a rough-and-ready recognition of private actors as invaluable stakeholders above and beyond states in standard-setting procedures? According to current usage, it could be both, either, or neither of the above. In fact, the concept of hybridity is rarely applied with any precision. In practice, it is generally employed to serve as a handy *default attribute* of the power of standards in global governance, part of a lexical register chosen to accommodate the multiple and contradictory understandings of their authority at the global level. After briefly reviewing how the notion of hybridity is used as a sort of "second-best" – or default – criterion, the following account suggests that the notion of hybridity, when used in the ordinary meaning of the word, is likely to reinforce – not qualify or clarify – the ambiguity of power relations more broadly involved in global governance. The argument is further supported by a detailed genealogy of the concept of governance which shows that it primarily enables authority to be exerted without full control of sovereign rights. While this sheds light on the common use of hybridity to deal with the ambiguous power of standards, it does not dissect the defining properties of such power. This will be done in the subsequent section focused on what hybrids eventually are or can be.

Many studies attempt to respond to the definitional challenge raised by the pervasive influence of standards and other kinds of market instruments in the functioning of the global economy by invoking the term 'hybrid'. Of note in this regard are debates in legal studies on complementarity and rivalry in the intersection of public and private standards, hard and soft law, and the layering of rules in 'old' and 'new' forms of governance, especially in the context of the European Union and the WTO (Mahler, 2007; Trubek and Trubek, 2007; Bartley, 2011; Zumbansen, 2011; Jurcys et al., 2013; Frydman, 2014; Pauwelyn, 2014). The notion of hybridity is brought in to characterise a distinct feature of regulation closer to a society-centred approach reaching out to a global level of analysis. According to Kjaer, the hybrid dimension of such a regulatory environment has even become 'common place insofar as the combined forces of globalisation and privatisation and an increased reliance on self-regulation have resulted in the emergence of regulatory arrangements which combine elements of several legal orders' (Kjaer, 2013: 3).[1] With the growth of third-party independent or semi-independent actors, standard-setting bodies and accreditation agencies, deregulation has indeed given way to the emergence, extension, and consolidation of new and more complex forms of regulation. This is all the more visible in the aftermath of the global financial crisis. Levi-Faur emphasises that this new golden age of regulation prompts a 'hybrid architecture of regulatory capitalism' (Levi-Faur, 2011a: 5). In today's world, a more comprehensive understanding of regulation should take stock of 'different systems of control, where statist regulation co-evolves with civil regulation, national regulation expands with international and global regulation, private regulation co-evolves and expands with public regulation, business regulation co-evolves with social regulation, voluntary regulations expand with coercive ones, and the market itself is used or mobilised as a regulatory mechanism' (Levi-Faur, 2011b: 668). In this analysis, hybridity describes the juxtaposition of state, market, and civil society actors in nearly all of the twenty-seven possible forms of regulatory design that combine regulators, regulatees, and third parties (Levi-Faur, 2011a: table 1, p.9). While this helps shed light on recent changes in the politics of regulation, the hybrid attribute is mostly used to denote the complexity that derives from the involvement of new actors in the regulatory design of capitalism, whether market-based or not-for-profit civil society organisations.

With the concept of 'innovation hybrids', Weiss brings the semantic field of hybridity one step closer to a major feature of contemporary global political economy. She draws upon Koppell's (2003) organisational typology of quasi-governments as complex partnership arrangements between public and private actors set in the distinct context of market organisation and innovation in the United States. Far from being confined to an organisational feature of an assumed neoliberal policy privileging privatisation and outsourcing, the concept of innovation hybrids carries, for Weiss, a much wider implication: they 'blunt the [national security] state's impact and blur its visible presence in economic governance, avoid political blockage, and promote the business of innovation' (Weiss, 2014: 147). The pervasiveness and significance of innovation hybrids is evidence of the extent to which technological pre-eminence has furthered the American model of capitalism and sustained American military dominance through the

[1] For the research agenda in global administrative law, see, among others, Cassese (2005) and Kingsbury et al. (2005).

dramatic changes in the security environment from the Cold War years to the post-9/11 era. This broader understanding of the hybrid attribute of contemporary global political economy arrangements echoes the argument put forward in Hurt and Lipschutz (2016). As stated in the introduction, here the hybrid rule constitutes a new phase of state formation. In contrast to analyses that emphasise how neoliberalism led to a retreat of the state in favour of privatisation, the authors maintain that state power is enhanced by privatisation and the ensuing depolitisation of the public sphere. As such, political developments often lack accountability; they can just as well help reinforce a closure of the public domain and accommodate a more authoritarian capitalist regime. New research on transitions has also acknowledged the importance of 'regime-hybridity' in developing countries to understand the role that democratic and undemocratic components of political regimes play in the trade-offs between formal and substantial democracy in economic transformation (Zinecker, 2009).

A prominent feature of contemporary global politics is indeed the ability of a wide range of agents to cooperate across borders to establish rules recognised as legitimate by states and non-state actors that have not formally delegated their sovereign rights for such mandates.[2] The scale at which globalisation is transforming the spatial organisation of social relations and production processes has magnified not only the way in which communities and issues are linked across nations, regions, and continents but also the power relations behind them. It is in this respect that international standards and global governance can be viewed as parts of a policy project supporting the involvement of new actors in the policy process, assuming that they would better tackle complex issues across borders. If left to either market self-regulation or plain state regulation, the argument goes, these transformations would be difficult to manage, lack efficiency, and, in the end, legitimacy. To be sure, cross-border modes of cooperation between public and private actors have not replaced the authority of the nation-state, and there is no reason to believe that they will in the near future. Yet, nation-states, as Held et al. (1999: 49) pointed out long ago, 'have gradually become enmeshed in and functionally part of a larger pattern of global transformations and

[2] It should be noted that private and informal ententes were also players in the game of organising capitalism at a time when the Westphalian interstate system supposedly conferred exclusive sovereign and territorial authority on states (Osiander, 2001; Teschke, 2002). Even at such a landmark moment in Westphalian history as the end of World War I, when the logic of colonies and empires started to give way to the universal interstate system and the principle of national self-determination, the diplomacy of war debts and reparation payments were mostly left in private hands (Hogan, 1977).

global flows'. If international policy coordination was hitherto chiefly played out within the confines of interstate multilateralism, and the power struggles therein, the project of global governance has taken a more depoliticised and functionalist turn. Where does this come from and how does it bring us back to the ubiquity of hybrids to which international standards belong?

At its core, the notion of governance refers to the act of exerting power without the appearance of doing so. As Guzzini points out, 'the two concepts of power and governance, although related, should not be conflated' (Guzzini, 2012: 3). Accordingly, the rise of global governance does not just reflect a diffusion of power; it also allows for informal rule that re-articulates the global political order and 'may well increase control' (Guzzini, 2012: 27). The notion of global governance thus echoes what we have seen so far with hybrids. Power mechanisms of any and all political, ideological, and technical persuasions are likely to take advantage of a fuzzy understanding of actors, forms, and sites associated with the exercise of authority. This suggests that the usage of the notion of governance enables the exercise of authority over a defined domain and population without the plain attributes of power imparted by sovereign rights. The following account looks at the genealogy of global governance from such a perspective focused on the ambiguous power relations highlighted by the notion of hybrid.[3]

First employed in the thirteenth century to denote the action or manner of governing, the term 'governance' initially referred simply to government. Yet, its reference object progressively moved away from the heart of power in the context of the development of the modern state, the centralisation of political authority, and the transformation of the principle of sovereignty that went together with the rise of modernity. The term eventually decoupled entirely from the actual centre of sovereign power. As Hewitt de Alcántara (1998: 109 – my translation) points out, 'while the concept applies to many situations where there is no political system as such, it still implies the existence of a political process'. Its usage has even come to presume governing practices that thoroughly exclude the type of political sovereignty found in modern democracies. As we will see later, this can be particularly advantageous in situations when sovereign states are confronted with groups that subscribe to different political orientations and push for different political

[3] A vast number of studies exist on the theories and conceptualisations of global governance; beyond those already quoted, see, in particular: (Hewson and Sinclair, 1999; Murphy, 2000; Paye, 2005; Payne and Phillips, 2014).

projects. In these cases, governance serves as a handy approach to reform in lieu of formal and more radical institutional change.

This understanding of governance has proved strikingly fruitful. The term came in political vogue at the turn of the twentieth century, in response to the first convulsions of independence within the British Empire. Not surprisingly, governance was used as a woolly definition of sovereign self-determination to justify reforms in the colonial status of key Crown territories, particularly India in the wake of the crisis related to the partition of Bengal (Silburn, 1910; Stuart-Linton, 1912; Low, 1913). A few decades later, pioneer studies in management and organisational studies reclaimed the notion in reference to an even more specific feature of economic power. Here, governance was seen as a tool to solve issues arising from the separation between capital management and capital ownership in the context of big firms. The relationship between the two dimensions of capital control, which had already been legally codified in the financial schemes of long distance trade established in the early days of mercantile capitalism, was further refined by the introduction of the concept of governance in the economics literature of the 1930s. Unlike state law, corporate governance focused on new models of corporate decision-making and behaviour in response to the rising power of waged managers in large American firms since the end of the nineteenth century. This analysis of the advent of the new managerial figure and the governance functions of corporate managers was introduced by Berle and Means (1932).

The notion of governance gathered steam in the context of the managerial revolution of Fordism and Keynesianism. It marked the successful rise to power of managers, who had started to share with the state and trade unions the responsibility for mass production, mass consumption, and the redistribution of high productivity gains. After several decades of undisputed consensus, the function of managers was again questioned at the beginning of the 1970s, first in the United States and then in Europe and the rest of the world. With falling productivity and increasing wage claims, trade-offs between owners and managers took a new turn – this time in favour of the former. New modelling techniques were produced, which reoriented the management utility function towards the valuation of shares on the stock market (Pérez, 2003: 35). Governance became a paragon of shareholders' value-based management and, at the same time, a watchword of the 1970s attempts to liberalise various aspects of the political order away from governmental or intergovernmental decision-making processes. With the demise of Fordism, owners of capital and top managers claimed the need to replace sovereign governments with corporate governance in arbitrating the distribution of productivity gains between capital and labour.

This shift went one step forward in the late 1980s. Governance, hitherto confined to the world of corporate management, was now to reach the four corners of the world. International financial institutions, such as the World Bank Group and the International Monetary Fund, greatly contributed to the popularisation of the concept of governance in reaction to mounting critiques over the repeated failures of development aid and structural adjustment programmes. With an emphasis on *good* governance, they recognised the need to complement purely market-oriented development models with measures to improve the quality of institutions in charge of reforms. Good governance enabled international institutions to abdicate responsibility for any development failure by adding domestic requirements as conditions for funding development programmes out of the debt crisis.[4] It followed, so the argument went, that the governments of incompetent and corrupt developing countries had to relinquish control to the external constraints of the world economy. To pick up on Gutner's taxicab analogy, international institutions used the language of good governance to issue explicit policy instructions while still leaving developing countries in the driving seat; *de jure* authority remained with the state, but *de facto* control shifted elsewhere (Gutner, 2010). Without let or hindrance, development finance made its way deep into the heart of the sovereign political sphere under the smoke and mirrors of good governance and new management principles (Osmont, 1998).

The establishment of the Commission on Global Governance in 1992, on the back of the post-Cold War burst of enthusiasm for greater collective responsibility, lent even more credit to the notion of global governance. However, its definition remained vague, as it describes 'the sum of the many ways individuals and institutions, public and private, manage their common affairs' (Commission on Global Governance, 1995: 2–3). Unsurprisingly, the work of the Commission did not prevent successive studies from casting doubt on the likelihood of governance undermining multilateralism as the prevailing form of collective action within the United Nations system. Perhaps de Senarclens put it best: 'advocates of this prescriptive approach tend to mingle all actors of the international realm in a large and woolly set, without hierarchy regarding their roles and political influence on systems of regulation [which results in the] naïve valorisation of non-state actors, particular multinational corporations, non-governmental organisations and international organisations' (Senarclens, 1999: 201). Behind the veil

[4] The rise and fall of the orthodox good governance doctrine among multilateral development agencies is well known. The landmark reports of the World Bank are the following: (World Bank, 1989, 1997, 2002).

of complexity and comprehensiveness, global governance is far from ideologically neutral. In fact, there is no scarcity of scholars describing it as an ideological project aimed at enforcing a particular – in most accounts, neoliberal – world order (Payne, 2005).

Governance, in one form or another, features equally heavily in the discourse around institutional innovation at the level of the European Union. Not only is the notion well suited to a supranational institution without full sovereign rights; it can also be used to sidestep the old debate between intergovernmentalists and neofunctionalists. 'Multi-level governance' made its debut as a European research programme in the 1990s (Marks et al., 1996; Hix, 1998; Tömmel and Verdun, 2009). The word now stands in the very title of the latest European Treaty that has come into force in the aftermath of the Euro crisis.[5] But, yet again, the undefined usage of this word leaves considerable leeway to private-public partnerships. It also offers no clear boundaries between, on the one hand, the political and administrative implications of the European tradition of continuous negotiation and, on the other, institutional developments taking place up and down centralised states. It is no coincidence that a great deal of discussion in the literature revolves around the extent to which the language of (multi-level) governance has taken over the language, if not the practice, of democracy, by gradually displacing the notions of expertise, representation, transparency, accountability, and legitimacy (Brassett and Tsingou, 2011; Keohane, 2011; Weiss, 2011). For a long time, it has also given cause for significant asymmetry between the degree of market discipline imposed by the Union on macroeconomic and monetary issues, and the limited room for manoeuvre left to member states for social and regulatory issues (Holman, 2004). More generally, the concept of governance has been at the core of the research programme on limited statehood, seeing the plain attributes of power imparted by sovereign rights as an exception of the Western modern nation-state rather than the rule (Risse et al., 2018).

Thus far, I showed how governance has meant quite different things throughout history, with increasing ambiguity regarding the attributes on which it lays claims to the exercise of authority. While the notion was confined to constituted powers in the *Ancien Régime*, the development of the territorial state and the rise of modern democracies progressively decoupled governance from government. Unlike the latter, governance refers to carrying out governing tasks without sovereign powers, generally

[5] The Treaty on Stability, Coordination and Governance was signed on 2 March 2012 by the leaders of all the then euro area members and eight other EU member states, and entered into force on 1 January 2013.

in new domains such as corporate management, development finance, and macroeconomic regulation. Governance has thus achieved prominence both as a normative compass for legitimating policies claimed to be closer to the people (i.e. good/democratic governance) and as an analytical tool to explore the emergence of a form of polity that is a step removed from both the state and the people (i.e. multi-level/network/informal/private governance). In both cases it remains ambiguous. So long as the notion of governance rests on a fuzzy definition of regulatory authority, qualifying it also as hybrid will hardly help shed light on the definitional criteria of its working institutions.

Since the turn of the millennium and the heydays of globalisation, defining governance, authority, or power as 'hybrid' became almost idiomatic in studies focused on the new patterns and actors of regulation involved in contemporary capitalism. Sassen, for example, stressed that 'the mix of processes we describe as globalization is indeed producing, deep inside the national state, a very partial but significant form of authority, a hybrid that is neither fully private nor fully public, neither fully national nor fully global' (Sassen, 2003b: 10). In a report of the French Conseil d'Analyse Économique, an institution under the aegis of the Prime Minister, 'hybrid governance' was considered the most appropriate way to frame the reform of the world order (Jacquet et al., 2002: 74–92). Similarly, the eminent development economist Gerald K. Helleiner (2001: 245) predicted that 'hybrid private–public arrangements' would probably count among the key purveyors of public goods at the global level. Two decades on, the catchword is still very much in the limelight. The number of actors and issues defined as hybrid has considerably increased. The range of fields concerned has also considerably expanded, with discourses on hybrid regulation featuring in sociology of organisations and production, public administration and administrative law, or security and warfare studies (see among others: Djelic and Quack, 2010; Miller et al., 2010; Acheraïou, 2011; Levi-Faur, 2011a; Belloni, 2012; Jurcys et al., 2013; Weiss, 2014; Hurt and Lipschutz, 2016; Leander, 2016; Lanoszka, 2016; Bair, 2017). According to Djelic and Quack (2010: 383) the emergence of transnational communities can thus be described as a permanent fixture of 'fluid … and hybrid formations out of formal organization and/or networks' supporting a new form of governance in a complex world. For their part, Dezalay and Garth maintain that global governance relies on competing forms of expertise on a transnational space; here, hybridity is seen as instrumental to overcoming the difficulties encountered in the recognition of this cognitive power in developing countries: a 'process of hybridization permits the progressive putting in place of new social usages built around foreign governance mechanisms' (Dezalay and Garth, 2011: 282) and presupposes the alteration of a prior 'logic that accounted for the coherence and relative efficacy of the governing device in the country of origin' (Dezalay and Garth, 2011: 277). All in all, hybridity comes through as a significant attribute of actors and practices involved in the increasingly complex process of globalisation. Yet, be it 'fluid' – as Djelic and Quack take it – or 'altered' – as Dezalay and Garth would put it – hybridity betrays more the lack of clear defining attributes than an effort to assign global governance a distinct feature.

Thus, returning to my question asking why the notion of hybrid is omnipresent in studies on standards and contemporary global governance debates, it appears particularly handy to take stock of the complexity instigated by new patterns of public and private regulation in contemporary capitalism. While mostly focused on the rise of private actors and standards, it entails much ambiguity on the defining criteria of supposedly new arrangements organising the world economy. It largely remains a *default attribute*. This second-best categorisation echoes the fuzziness of the concept of governance itself, and is likely to reinforce it. Far from being a mere non-sense, hybridity helps to accommodate multiple and contradictory understandings of global governance. Even if left without further specification, the integration of multiple and contradictory meanings and practices is thus not at all pointless. As Acheraïou points out, the protean nature of discourses on hybridity reflects a 'structural flexibility', which greatly contributes to its resilience; it integrates multiple, contradictory, let alone irreconcilable, lines of reasoning and 'lends itself to ready appropriation by almost anyone, to serve almost any political or ideological purpose' (Acheraïou, 2011: 153). Moreover, the vagueness of the notion is likely to support political disengagement and historical shortsightedness, thereby leaving the door open for misappropriation by both dominant and contending forces. Yet, the concept remains dubious if employed without further historical, geographical, and conceptual roots. This brings me to my second set of questions: can the term hybrid be employed as something more than a default attribute? What are its defining properties, and how does it reflect the distinct, non-conventional form of market creation and regulation embodied by the widespread use of standards in the international economy?

What Hybrids Are

Not just a default attribute, the notion of hybrid conveys *substantive attributes* which can help make sense of standards as non-conventional forms of power in contemporary capitalism. From this angle, ambiguity becomes a prevailing feature of the criteria that define the authority of standards in creating and regulating markets. Yet, far from a vague conglomeration of actors involved in setting complex rules, ambiguity is seen here as an ontological property of the new tools of global governance of which international standards are a case in point.

In order to understand this ontology, I draw on insights from semiotics, sociology of science, technology and society, and post-colonial studies. In the previous section I showed that the default position of studies laying emphasis on the notion of hybrid would just get us to describe the intermingling of public and private standards as a juxtaposition of state, market, and civil society actors. Semiotics helps us to see hybridity as much more than a default attribute, with a rich and long etymology; its ambiguous meaning and ambivalent values reach back to early representations of collective life. Studies in the sociology of science, technology, and society, for their part, open our eyes beyond the private-public nexus of hybrid governance debates. They emphasise the ambiguous relationship between nature and society, leading to the understanding that standards are never mere technical specifications and always convey social values, be they implicit. Finally, post-colonial studies of hybridity lend support to understanding the cultural and spatial underpinning of power relations conveyed by the ability of standards to extend their authority beyond borders.

First of all, in order to fully grasp the defining properties of hybrids, it is worth thinking back to ancestral figures of human imaginary. From such a larger semiotic perspective, hybrids pervade all sorts of myths' narratives across time and space (Uranie, 1996). Contrary to the ideal of unity and simplicity found in classicism, they gain their persuasive power through fabulous and multifaceted dimensions. Hybrid creatures form powerful legendary wholes; even if each of their parts is of real and well-defined origins. In Ancient Greece and Rome, couplings between humans and animals usually gave birth to malign monsters. Philologically, the orthography of the word 'hybrid' was quickly twisted to express more fully the awe conveyed by such creatures. In Latin, *ibrida* was used by the Roman naturalist Pliny to describe the crossbreeding of a sow with a wild boar. Shortly afterwards, it became *hybridia* – replacing the first *i* with *y* was meant to call to mind the Greek word *hybris*, which connotes all sorts of excess and transgression, possibly leading to violence. For instance, the fire-breathing Chimera was an awe-inspiring creature able to melt, devour, and vomit anybody and anything she met; for centuries, she personified evil in early Christian art (Godin, 1996: 46).[6] The

[6] Reference to the figure of the chimera is also made by Leander (2014) to analyse the enmeshed and elusive characteristics of the public-private divide at the core of US National Inteligence and security governance.

Minotaur, the monster with a bull's head and a human body, is probably the most famous of those malign creatures; living in the centre of the Cretan Labyrinth, every ninth year he devoured seven youths and seven maiden sent from Athens as tribute. When the coupling is between humans and gods, however, hybrids usually take a much more benign shape. In those cases, 'the contribution of divine blood is like a regeneration of the human race' (Brémond, 1996). Here, the figure of Helen can jog our memory. Daughter of the mortal Leda, wife of King Tyndareus of Sparta, yet fathered by Zeus, she is remembered not only for having been the most beautiful mortal on earth, but also for prompting the Trojan War, the founding moment of Greek civilisation. Hence, from a semiotic perspective, rather than being entertaining and naïve characters, hybrid figures disclose fundamental features of the organisation of collective life. Their power lies on a juxtaposition of life forms and qualities that transcends singular purposes. Ambiguity is in itself central to understanding this form of power. As Godin (1996: 40) emphasises, the basic forces of hybrids at work are their *ontological ambiguity*, wavering between reality and the imaginary, and their *affective ambivalence*, hesitating between repulsion and seduction. In other words, it is by means of such ontological ambiguity and affective ambivalence that 'the hybrid has the power of the Whole that s/he symbolizes' (Godin, 1996: 43). In the present world, this underlines how governance instruments such as international standards draw their power from the ambiguity that characterises them as new tools made up from easily identified parts but assembled in such a way as to form inventive artefacts in charge of shaping significant transformations of contemporary capitalism. Similarly, ambivalent views call up feelings of both attraction and repulsion towards such instruments closely related to opportunities and threats associated with globalisation – largely depending on where one stands on the scale of the multifaceted hierarchy of global capitalism.

Studies in the sociology of science, technology and society situate hybrids in the broader context of modernity and the major debates on its crisis. In his analysis of the rise of the 'risk society', Beck ties the notion of risk to the ways a 'hybrid society watches, describes, values and criticizes its own hybridity' (Beck, 1992, 2000: 221). Our society is thus understood to become intrinsically reflexive. Technical issues, previously confined to private choices or narrow bureaucratic decision-making, are increasingly opened to democratic scrutiny. In underlining the growing significance of public scientific and political controversies, Beck considers that 'the notion of a "hybrid"' world is necessary, but insufficient [since it] says what it is not – not nature and not society etc. – but it does not really say what it is' (Beck, 2000: 221). He unmistakably recognises the significance of Latour's critique of modernity in this discussion. Yet he keeps seeing it as more of a negative than a positive concept and, therefore, failing to provide much guidance to further enquiry.

This is quite misleading. 'Hybrid' is, in fact, the conceptual linchpin of Latour's analysis of the current crisis of modernity and of the theory built to overcome the conundrum that follows from its dichotomic framework of thought (Latour, 1991). With examples drawn from the daily reading of the newspaper, such as the hole in the ozone layer over the Antarctic, HIV contaminations, or competition over computer microchips, hybrids refer to what Latour calls 'quasi-objects' belonging neither to nature nor to society but to both of them. Rather than merely negative, the concept reflects an attempt to reunify the understanding of a world torn apart by the advent of modernity. While this combination of natural properties and cultural traits was considered self-evident to pre-modern societies, Latour suggests that the great illusion of modernity – leading to its current crisis – is the belief in the ability to conceal this co-constitution of nature and culture. In its grand design of purification and separation, the argument goes, modernity made nature and society opposite poles in the organisation of collective life. Not only did this process liberate forces to dominate nature in a limitless perspective but it also initiated the domination by the West of the rest of the world, identified as lacking the modern scientific knowledge required to avoid being blinded by the confusion between sign and thing (Latour, 1991: 135). From this viewpoint, hybrids reflect the erosion of the great divide that modernity failed to establish between society and nature, humans and non-humans, society and science. With many controversies sketching out imbroglios of scientific, political, economic, legal, and other concerns, standards codify not only technical specifications but more broadly the proliferation of 'quasi-objects'. Such hybrids call for a complete shift of perspective, which amounts to the application of a principle of symmetry to the analysis of the co-constitutive properties of nature and society. This approach, commonly known as Actor–Network Theory, analyses those co-constitutive properties as networks. Their productive tension overcomes the opposition between structures and agents through processes that can be traced at various scales and across different spaces.[7] This is the purpose of Latour's so-called new Constitution, whose first guarantee is to avoid separating nature and society again: 'nature and society are not two distinct poles, but one and

[7] For an early constructive critique of the relevance of Latour's theory for international relations, see (Elam, 1999); there has been a flurry of scholarship transposing Actor–Network Theory into the field of international relations over the last few years. See, among others (Barry, 2013; Best and Walters, 2013).

the same production of successive states of societies-natures' (Latour, 1991: 191). While Latour thinks of 'Parliaments of Things' as institutional extensions of this line of thought, Callon et al. conceive 'hybrid fora' as the new arenas where expert knowledge mingles with lay knowledge to arrive at technical choices involving the wider public and thus encourage a shift from delegative to dialogical democracy (Callon et al., 2001: 189). Such embeddedness of scientific and technological choices into social, cultural, and political contexts and institutions can also be conceived as a co-production, according to which 'there cannot be a proper history of scientific things independent of power and culture' (Jasanoff, 2004: 21). A number of arenas have put into operation this new form of regulatory politics, such as nuclear waste management, the fight against AIDS epidemics, GMO technologies, and nanotechnologies. For my part, I designed the platform INTERNORM, a pilot project funded by the University of Lausanne (2010–2014) to foster the involvement of civil society associations in ISO technical committees. That interactive knowledge centre based on the sharing of academic skills, ad-hoc expertise, and the experiences accumulated by consumer associations, environmental associations, and trade unions gave a unique opportunity to experience a new way of responding to the democratic deficit rampant in the field of standardisation.[8]

Thus, for scholars of science and technology studies, the hybridity of artefacts such as standards is much more than a default attribute describing heterogeneous developments in contemporary global governance. Rather than a mere juxtaposition of private and public actors, hybridity reflects a fundamental property of our relation to the world, in which the two poles of nature and society are intrinsically co-constituted. Hybridity embraces dual function as a form of authority that is used to govern but also as a potential means to engage and resist such forms of governing – all the more so when hybrid fora increasingly shape the organisation of markets (Callon et al., 2002: footnote 11). Theorists sharing this alternative view consider that nature–society relationships gain, above all, a foothold in local and ad-hoc conditions, be it the laboratory of Louis Pasteur (Latour, 1984), the outpatient department of a large hospital, or so-called publifora, where assemblies of citizens debate new technological challenges (Callon et al., 2001, chap. 5). While this helps to understand global governance as a socially embedded practice in line with the so-called practice turn in IR theory (Best, 2013: 22–25), it leaves us short of a proper understanding of the transnational nature and the

[8] See: Hauert et al. (2016) and Graz and Hauert (2019).

global reach of many institutional arrangements involved in such practices. Moreover, such an emphasis on network process tracing of quasi-objects leaves little space for the wide range of actors involved in such contexts, their strategies and capacity to act – and still less for any consideration of the deterritorialisation of sovereign space and power as conventionally understood. This prompts us to pay attention to the third strand of scholarship that has given hybrids central stage.

Post-colonial studies view hybridity as a result, not of mingling nature and society, but of the influence of colonialism on the blurred subjectivities and identities of the colonised subjects throughout the period of decolonisation as well as globalisation. The concept was arguably formulated by Bhabha (1994) and subsequently widely discussed by authors in literary criticism such as Lionnet (1995) and Young (1995). To some extent, this debate alone gave a fundamental contribution to the emergence of the very field of post-colonial studies.

Bhabha drew the concept from the Russian linguist Bakhtin in order to overcome Saïd's (1978) analysis of orientalism, which was blamed for its totalising view of power and colonial discursive practices. According to Bakhtin, languages evolve in society like 'unintentional hybrids', with mixed worldviews remaining 'mute and opaque' (Bakhtin, 1981: 360). In contrast, the reconstruction of language by a novelist is often an 'intentional and conscious hybrid' (Bakhtin, 1981: 366), an 'artistically organized system for bringing different languages in contact with one another' (Bakhtin, 1981: 361). In coining the concept of hybridity, Bhabha makes a similar argument about the ability to transform an unintentional condition of dominance into an intentional strategy of emancipation. On that account, 'moments of hybridity' become moments of 'historical agency' (Bhabha, 1994: 208). According to Young, the significance of this argument is stupendous: 'By grafting the Bhaktinian notion of the subversive and dialogical force of hybridity onto the ambivalence in the colonial encounter, ... Bhabha has shifted this subversion of authority through hybridization to the dialogical situation of colonialism' (Young, 1995: 22). In doing so, post-colonial studies do not merely define hybridity as a rejection of binary relations of power. They lay emphasis on alternative concepts, such as difference, multiplicity, plurality, fluidity, and ambivalence, in order to stress how existing situations of colonial domination can and have become instruments of resistance. For instance, Ní Mhurchú draws on hybridity to describe certain experiences of subjectivity 'as a form of ambiguity within, rather than a form of presence across, several nations ... at the intersection of citizenship and migration' (Ní Mhurchú, 2015: 167). A number of studies question, however, such possibility of dismantling power structures and idealised valorisations of the struggles of subaltern subjects. According to Prabhu, there are serious doubts that hybridity holds for a "radical conception of agency" if left to such flattened and fanciful means of resistance (Prabhu, 2007: 2). Together with Prabhu, Kraidy and Acheraïou have also thoroughly criticised the inability of the post-colonial understanding of hybridity to properly address the question of agency in relation to the material structures of power in globalisation, let alone its tacit complicity with a Eurocentric post-modern ethos (Kraidy, 2005; Acheraïou, 2011).

Despite these criticisms, dismissing the concept of hybridity out of hand for our understanding of the power of standards would be tantamount to throwing the baby out with the bath water. Besides science and technology studies which provide new insights into the significance of 'quasi-objects' mediating nature-society relations, the cultural and inter-subjective processes that inscribe the spatial domination of globalisation on a concrete basis are the most obvious lessons to be drawn from post-colonial studies. This is why, for instance, research on the degrading and over-standardised labour of call centres in India and elsewhere, as well as other predicaments associated with the growth of remote services enabled by information and communication technologies, draw so easily on the post-colonial conceptual toolkit (Das and Dharwadkar, 2009). From a wider and deeper socio-historical perspective, Acheraïou makes a similar point on the power dynamics and multifarious nature of hybridity in ancient empires, characterised by a spatial domination which was heavily dependent on administrative and political syncretism. Transposed to our contemporary context, the concept of hybridity can thus help us to describe such 'syncretic modes of governance' (Acheraïou, 2011: 19) or what Cox describes as a 'plural world of coexisting civilisations' (Cox, 2002: 56). Besides shedding light on how standards can be seen as operational devices used to create and regulate markets with strong cultural underpinnings, post-colonial approaches can also reveal concern towards disaggregated spatial structures. Countless studies have focused on the 'interrelated, if not overlapping, spaces' of *métissage* and diaspora (Lionnet, 1995: 7). The lexical reference to hybridity here helps us to take on an equally illusory understanding of formal territorial sovereign space, still closely or loosely shared among IR scholars. To some extent, it echoes van der Pilj's far-reaching critique of IR theory as unable to account for various 'modes of foreign relations', in which human communities combine different ways to occupy space, to secure it, and to organise exchange between each other. While the global governance project lies on a formal equality of sovereign states, it not only brings on-board 'the exploitation of nature and society on a world scale' but also continues to coexist with primeval nomad and imperial modes (Pijl, 2014: viii). Ultimately, the importance given to standards for market access across borders entails a hybrid space whose transnational logic overlaps in many ways states' territorial sovereignty. Being hybrid, however, this form of authority not only governs market but can also be seen from its opposite side, as a resistance to market power.

In summary, semiotics, science, technology, and society studies, as well as post-colonial approaches, provide insights into the many ways in which hybridity has intrinsic properties. All of them can help define the issues at stake, the actors involved, and the cultural and spatial environment where new actors claim a say in the global governance to which standardisation bodies belong. Just as the ISO is made up of more than 150 private and public bodies designed as the 'most representative of standardisation in their country', many other standard development organisations exist and are broadly recognised so long as their instruments are adopted by markets. Quality management standards such as Six Sigma, multi-stakeholders' initiatives like the Roundtable on Sustainable Palm Oil, and metrics designed for professional practices by well-organised associations such as the International Standards on Auditing (ISA) produced by the International Auditing and Assurance Standards Board (IAASB) of the International Federation of Accountants are all examples of specifications that confer authority (although with very disparate reference bases) on actors seeking market access across borders. Hybridity describes such an ambiguous juxtaposition of power instances that intermingle with the interstate system. As with hybrids, this phenomenon is essentially Janus-faced. Yet, in addition to the resistance strategy envisioned by post-colonial scholarship or networks of symmetric mediations between nature and society, as conceived by Actor–Network Theory, the phenomenon remains anchored in asymmetries of material power that support the global governance project of unifying markets across sovereign territories. In other words, contemporary hybrids sanction new objects and agents which, though fully real, are aggregated in such a way that their defining criteria entail inherent ambiguity and inherent ambivalence in their powers of attraction and identification. This brings us to our third argument, where we will try to specify the categories that remain ambiguous in the idea of hybridity.

How Hybrids Work

It would be presumptuous to reduce the hybrid power of standards in the global political economy to the ambiguous criteria conferring authority to their cross-border endeavour to create and regulate markets. New forms of collective action and international authority – as well as the broader and more disparate influence of non-state actors on the world stage – should be seen as a multi-dimensional phenomenon with multiple material and symbolic aspects. Responding to the question of who standardises what and where thus supposes an understanding of how ambiguity confers authority to new actors and new issues across sovereign spaces.

In her pioneer investigations on the *Retreat of the State*, Susan Strange suggested that 'between the two extremes of non-state authorities welcomed and opposed by states lie certain non-state authorities whose relation to governments is variable or ambiguous' (Strange, 1996: 94). In her concluding remarks, she notoriously equated the advent of non-state actors in the arena of global politics to Pinocchio's problem: at a loss when caught without any more strings to guide him. The lack of a clear definition of non-state actors in world politics has led, in her words, to 'a ramshackle assembly of conflicting sources of authority', making it particularly hard to decide 'where do allegiance, loyalty, identity lie' (Strange, 1996: 199). According to Cutler, it is precisely these conflicting sources of authority that create a new form of private authority in international affairs. Cutler emphasises in particular the political significance of legal doctrines that have twisted the status of the subject of law: 'the implication of treating corporations and individuals as objects and not subjects are deeply troubling empirically and normatively ... [W]hile transnational corporations and private business associations may be objects of law (*de jure*), they are in fact, operating as subjects (*de facto*)' (Cutler, 1999, 2003: 149). Analyses in terms of private international authority shed light on the range of actors to have gained authority in an international context that traditionally denied them that privilege. They paved the way for in-depth studies of firms and inter-firm cooperation leading to political roles for actors traditionally associated with the private sphere of economic transactions. They also raised the troubling normative implications of an authority geared towards maximising capital gains and concealing the instruments serving those ends (Gill and Cutler, 2015). Yet, focused on the cooperation of firms across borders, this approach remains primarily concerned with a sub-set of actors. Since then, countless studies have been published on the wide range of political positions vis-à-vis global governance issues taken on by other non-state actors, such as non-governmental organisations, social movements, global civil society platforms, and, not least, transnational criminal organisations. From technical self-regulation to corporate social responsibility, from environment and labour standards to financial and accounting rules, much of the literature is focused on who governs the global economy through private regulatory tools (Hall and Biersteker, 2002b; Schirm, 2004; Grande and Pauly, 2005; Krause Hansen, 2008; Avant et al., 2010; Djelic and Quack,

2010; Green, 2014; Payne and Phillips, 2014; Gill and Cutler, 2015). There can be sharp disagreement as to the meaning attributed to the prominence of non-state actors, variously understood as suppliers of private standards making up for the failure of governments to embrace such tasks or as influential corporate actors shaping regulatory outcomes in favour of the financialisation of global capitalism. With a focus on voluntary standards as privileged instruments of global governance mechanisms situated somewhere between those two poles of public and private power, this book aims at looking not only at the ability of private actors and civil society organisations to shape global regulation across borders. Two other aspects play a key role in the reconfiguration of global capitalism: one is the scope of practices involved in standardisation and the other is the reconfiguration of the spatial structure in which those practices are recognised and implemented to such an extent that compliance can be assumed on a transnational basis.

Casting the nature and the implications of the rise of hybrid authority across borders in a broader context thus requires us to consider and aggregate three distinct categories: the *subjects* wielding authority, the *objects* concerned, and the *space* of their deployment. For instance, international standards set by the ISO as well as those provided by the not-for-profit body ASTM International (originally known as American Society for Testing and Materials) entail numerous technical experts and national delegates who play the role of new actors in the nascent technical diplomacy world. The same experts and delegates also define the nature of the objects concerned (from nuts and bolts to sustainable innovation and societal responsibility) and the spatial structure in which they exert their power (on a national, regional, or global scale). The point here is to suggest that ambiguity not only defines the status of the actors involved in standardisation and regulation but also the scope of issues concerned and the space on which such authority is recognised. Following on from the introduction of this chapter, this non-conventional form of power is what I call a *transnational hybrid authority*, i.e. a form of authority based on the ambiguous juxtaposition of instances of power transforming the relation between transnational capitalism and territorial sovereignty. That said, I still need to specify those ambiguous categories that confer authority to new actors and new issues across sovereign spaces. Transnational hybrid authority is shaped by each of these three dimensions (actors, objects, space) as follows.

Hybrid Actors

The first dimension concerns the actors setting technical specifications, regulation mechanisms, and broader governance standards, plus the distinction between the private and the public spheres in which they operate. Many discussions on non-state authority and global governance focus on what has been referred to elsewhere as a 'diffusion of authority' (Guzzini and Neumann, 2012). As Payne and Philips point out, certain developments over the last two decades have led to a situation in which, 'as well as being pushed downwards, upwards and outwards to different spatial levels, authority and agency have dispersed to a wide range of actors at and across all these levels' (Payne and Phillips, 2014: 6). A proper understanding of the wide variety of actors in a position to set standards and shape regulation across borders depends to a great extent on the definition of the private/public distinction, and its interplay with the civil society. Despite variations between societies, the separation between the modern state and the economy has shaped social relations by distinguishing between the private and the public spheres (Cutler, 2003: 141–179). They remain closely related, reflecting two sides of the same coin. While the public sphere confers universal rights in the political domain, the private sphere brings them into play for the purpose of providing contractual rights in the economic and civil domains. The range of private actors claiming authority in international affairs is thus larger than what we refer to as the 'private sector' in narrow economic terms. It may include non-state actors such as trade unions, activist groups, women's organisations, professional associations, cadres and experts organised in ad-hoc bodies, advocacy or policy networks, elite clubs, and religious groups. This implies that the 'private' authority of non-state actors in international relations potentially includes any collective actor organised through formal or informal contractual relations within the realm of civil society. As Colàs (2002: 23) argues, civil society should not be viewed as a benign sphere of collective action outside the state system, but rather as a 'space of contested power relations where clashing interests play themselves out through analogous but unequal modes of collective agency'. Such contests may assign authority to some actors while undermining the authority of others.

Against this backdrop, hybrid actors ready to set standards span an *institutional continuum* whose poles include both the public and the private sphere. Ambiguity as to where actors should be situated on the continuum makes it possible to confer authority on those who were traditionally denied such privileges at the international level. It is indeed the lumping together of private and public attributes that bestows authority onto a set of actors who previously lacked the qualification to do so. In this view, hybrid actors look like the new wholes of ancient mythological figures: they are able to transcend the attributes of each of their former conditions. For instance, as Chapter 5 will show in full detail, the shift towards principles-based regulation, the use of internal models of solvency capital requirement, and qualitative requirements such as the Own Risk and Solvency Assessment (ORSA) have recently put private insurers in a position to set standards on their own and shape regulation in their own favour across borders. In so doing, governmental and intergovernmental regulatory bodies support and fully recognise the self-regulatory power of private insurers.

Whether we speak of non-state actors, private authority, or non-conventional forms of sovereignty and governance, two distinct conditions must be met for such hybrid forms of authority to be effective: the consent of actors who are subject to the rules without having been involved in their making; and explicit or implicit support by the state. Indeed, consent – rather than enforcement or explicit compliance – is a vital element in such configurations of power. As Cutler et al. (1999: 19) emphasise, 'those subject to the rules and decisions being made by private sector actors must accept them as legitimate, as the representations of experts and those "in authority"'. In the same vein, Djelic and Sahlin-Andersson (2006: 23) consider that non-state authorities enabling various forms of transnational governance hinge upon 'powerful institutional forces that altogether constitute a transnational culture or meaning system'. Similarly, Büthe emphasises the socio-political incentives for consent to such rules when private regulators are well aware that they 'may be more efficient if their rules (or at least their rule-making) are perceived as legitimate' (Büthe, 2010: 20). This aspect is closely related to the power of reputation in compliance processes, which substitutes command-and-control, hierarchical, and formal state regulation for informal and non-hierarchical governance. It is important, however, not to overemphasise this consensual underpinning of non-state authority, which I identified elsewhere as one among other limits of transnational private governance (Graz and Nölke, 2008).

As briefly stated in the introductory chapter, beyond this consensual dimension underlying the transnational hybrid authority of standards, states retain a central role in the rise of such authority. There is no consensus on how to conceptualise state recognition. But there is little disagreement on the overall complementary and subsidiary role of private actors in regard to state functions. As Payne and Philips point out, 'most complex governance arenas inevitably require that both types of actor are comprehensively engaged if progress towards solutions to policy dilemmas is to be achieved' (Payne and Phillips, 2014: 475). Similarly, Pauly and Grande underscore that 'the idea of reconstituting public authority implicates the institutional form of state sovereignty, and its scope and content as well' (Grande and Pauly, 2005: 16). Recent transformations in the sovereignty of states thus fully bring on board new agents of global governance to whom standard-setting bodies are only one among many examples. The analysis of global politics and non-state authority should therefore be 'not about the type of the agent, but the *character of the relationships*, both among governors and between governors and governed' (Avant et al., 2010: 3). This reflects what other scholars describe as a 're-articulation of regulatory authority', in which 'public regulation has indeed retreated in some areas of the economy, but in the same time other forms of governmental and inter-governmental regulation are actually being strengthened', such as in intellectual property rights, trade and investment, and humanitarian law (Utting, 2008; Ponte et al., 2011: 7).

To sum up, the transnational hybrid authority conferred on standard setters hinges on an institutional continuum marked off by the confines of the public and private sphere. Governments and intergovernmental institutions often support and fully recognise the power of non-state actors to set standards subsequently accepted by a wide range of actors not involved in their making. They are consequently likely to enhance their legitimacy.

Hybrid Objects

The second axis along which the power of standards is to be analysed are the objects on which they exercise their hybrid authority. Whereas the private/public nexus of the actors involved in setting standards can be located on an institutional continuum, this second dimension maps out a *material continuum* delineating what can be specified, standardised, and more generally regulated from the two opposing poles of the physical and societal world. Globalisation can hardly be dissociated from the impact and pace of technological innovation in such diverse industries as neuroscience and bioinformatics, knowledge process outsourcing of services supported by information and communication technologies, and big data behind individual hyper-connected services. The scope of technological change not only generates potential limits of a science-based economy against an allegedly given, natural environment. It also betrays a lack of democratic control over the proper use of technology in society. As mentioned previously, social studies of science and technology describe hybrids as quasi-objects across the divide established by modernity between nature, science, and society. The distinct hybridity of issues concerned with standardisation practices in this context emphasises the ambiguity of technological choices and innovation embedded in constellations of power and political struggles. In aggregating the relationship between human beings and nature, a transnational hybrid authority ranges from natural and invariable physical measures to constructed and historically bounded societal values. The extension of the scope of international standardisation provides ample evidence of the significance of this second dimension. For decades, standards were mostly confined to 'physical' standards, such as the size of screw threads, the resistance of materials, and units of measurement. They are now covering more and more 'societal' topics. Corporate standards in social responsibility (ISO 26000), risk management (ISO 31000), energy management (ISO 51000), or anti-bribery management (ISO 37000) are emblematic in this regard. Yet, even standards seen as highly technical, such as those developed by the ISO Technical Committee 229 on nanotechnologies, may include deep societal issues, such as occupational risk management for laboratories dealing with nanomaterials, safety data sheets used by workers in the preparation of nanomaterials, or the labelling of products containing nanomaterials likely to be bought by consumers.[9] In contrast to the widely held belief that complex technology could justify keeping democratic principles at bay, the hybrid premise highlights the fact that technology remains inextricably linked to society as a whole.

Political institutions often appear at a loss when facing the hybrid nature of the issues involved. Their complexity commonly justifies claims of experts to have a hold on it; their societal underpinning would, on the contrary, deny to expert technical committees the right to reach any final decision. And yet, hardly any decision can be reached today without some sort of scientific assessment, forecasting, or approval. Following a Weberian view on modern state power, this substitutes, to some extent, the role of bureaucracies in the foundations of authority and the domination of modern states in the structure of capitalism. Such 'technical authority' (Porter, 2005; Best, 2012) rests on a sustained ambiguity between technical and societal issues. Such hybrid status brings me back to the fundamental question of the relationship of human beings with nature. On analytical grounds, this ambiguity draws our attention to how the material continuum of our conceptual framework can link the opposing poles of the physical and societal world of standards – or what various authors have referred to as the 'Parliament of things' (Feenberg, 1991; Latour, 1991; Salomon, 1992). This supposes piecing together again the puzzle of what Latour defined as the quasi-objects that conceal the divide between the human and non-human world. Be they apparently more

[9] I draw this from the findings of the aforementioned INTERNORM project, funded by the University of Lausanne and devised to support the direct participation of civil society organisations in arenas setting international standards; for further information see: www.unil.ch/vei/internorm (accessed 24 August 2017).

technical, as in the field of nanotechnology, or more societal and cultural like all those dealing with social responsibility, the set of issues concerned by standards discussed earlier nevertheless link together societal stakes of collective life with its material, natural, and, more generally, physical dimensions. The domains concerned remain ambiguous. They tend to conceal the difference between an authority founded on scientific knowledge and technical expertise, and an authority built upon a formal mandate and with procedures in place for delegating the sovereign power formally conferred on individuals in democratic societies. As we will now see, the territorial basis of the state and the structural power of governments and markets remain beyond most forms of non-state authority. The ambiguous authority of standards here comes from their ability to be recognised across sovereign territories.

Hybrid Spaces

The third dimension of my analytical framework deals with the production of the space through which transnational hybrid authority is diffused and recognised across sovereign states. The question goes back to the central issue of how standards can ever be effective and how we comply with them. A whole host of organisations contribute the ways in which standards are used to shape global governance practices. Standard-setting bodies must have certain defining characteristics to be recognised as sufficiently legitimate to set specifications that cannot be seen as only technical. For instance, as Chapter 4 will show in detail, the new generation of mega-trade deals such as the Canada-European Union Comprehensive Economic and Trade Agreement (CETA) specifically target convergence in regulatory approaches and harmonisation of standards as one of their most prominent outcomes. The intertwinement of standards in regulatory enactments, contractual relations, and potential liabilities to public courts and private arbitration panels points here to a fragmentation of the political space that is distinct from the modern inside/outside divide that is supposed to shape the confines of autonomous political units commanding final authority within a defined sovereign territory.

Conventional answers to the question of why we obey standards set as rules not issued by states usually fall under two types. Materialist explanations focus either, from a neo-Marxist perspective, on the structural power of capital to impose market discipline or, from a more liberal approach, on the market structures (such as oligopolies, network effects, or the clout of reputation) that enable compliance with private regulations. Institutional and normative analyses focus for their part on the legitimacy puzzle of new governance arrangements, in particular on the significance of, and relation between, input legitimacy (the type of participation in decision-making), output legitimacy (the effectiveness, efficiency, and social justice of such rules), and throughput legitimacy (the quality of the deliberation processes to attain such rules) (Cutler, 2010). Here I take a different approach. Based on an evolutionary understanding of the relationship between space and power, I see compliance with non-state rules as resulting from the fragmentation of the political space, whereby its underlying logic is reorganised in a way that encroaches upon conflicting sources of transnational authority. The growing significance of non-state rules prompts states to adapt to the changing conditions of their environment in cumulative processes with irreversible consequences. According to the processual and sequential analysis of the changing institutions of capitalism pioneered by Veblen, this presumes a 'cumulative change, realized to be self-continuing or self-propagating and [with] no final term' (Veblen, 1919: 37, see also: Dopfer, 2005). Transposed to our present concerns, this echoes the idea according to which the 'territorial trap' of international political economy can only be overcome by 'showing how the domestic and the foreign come together under different historical circumstances rather than separating them into permanent opposition' (Agnew, 1994: 67).

The ambiguity of the relations within which standards are recognised and made effective spans a *spatial continuum* across multiple jurisdictions. This third axis of our analytical framework conflates two interlocking logics: the endogenous logic of territorial sovereignty, on the one hand, and the exogenous logic reinforcing the transnational underpinning of capitalism on the other. The idea of a dimension of continuity in the political space of modern nation-states goes against a conventional reading of globalisation as either a continuation of state sovereignty by other means or a sheer deterritorialisation of the nation-state. According to a critical approach that considers space as an output of social relations rather than physical design, the assumed spatial correlation between the nation and the state has never existed; neither have distinct spaces separating political entities and discrete domestic national economies (Osiander, 2001; Teschke, 2002; Cameron and Palan, 2003). As Nederveen Pieterse (2001) has noted, hybridity can be viewed from this perspective as particularly significant insofar as it problematises boundaries. If you think of the International Financial Reporting Standards set by the International Accounting Standards Board, they are typically seen as the outcome of the private expert authority embodied in professional accountancy bodies with strong British and European backgrounds. Their adoption by some 120 countries since 2001 (including all but five members of the G20) rests, however, on a complex mixture of private and public organisations that have endorsed them in one way or another. On the private side of the market for professional services that spans existing boundaries, the Big Four accounting firms (PWC, Ernst & Young, Deloitte, and KPMG) have, for instance, gained significant power as a result not only of their ability to shape standards but also of the expertise required to interpret them and provide professional judgement in their role of auditors. On the public side of sovereign spaces, national, regional, and international regulators have been instrumental in the swift spread of their adoption. The official support of the International Organisation of Securities Commissions (IOSCO), the references included in the Reports on Standards and Codes of the World Bank, the emphasis placed by the G20 since the global financial crisis and, *last but not least*, the European Regulation ((EC) No 1606/2002) that compelled the use of IFRS for all listed companies across the European Union as of 2005 are only the most significant examples of the role played by public regulators with greater or lesser connection with territorial sovereignty (Humphrey and Loft, 2011; Ramirez, 2013). This is how I have come to view the expansion of the spatial reach of standards as reflecting a transnational hybrid authority that occupies the cracks between the all-encompassing rules governing the global market and the enduring exclusive principle of territorial sovereignty.

How do the entwined exogenous and endogenous poles of this spatial continuum help us to obey standards not necessarily set by States? According to Palan, the institution of sovereignty carries out two closely related functions: 'the juridical expression of the principle that divides the planet into clearly demarcated lines of authority and responsibility' on the one hand and, on the other, 'the foundation of the national and international law of contract' required by capitalism (Palan, 2003: 86). From this view, standards need hybrid spaces that reinforce such dual nature of sovereignty. Their recognition rests on the territorial inscription of sovereignty in the same time as on the transnational guarantees given to the principle of contract inviolability in a world of globalised capitalism. Power mechanisms rooted on the territorial dimension tend to rely on an endogenous logic of recognition. They rely on social forces located within the territorial space of state sovereignty. Conversely, contractual market guarantees provided for the spatial expansion of capitalism across sovereign spaces convey exogenous forces. Compliance with standards rests in many ways on the ambiguity of a similar dual mechanism. From an endogenous logic, what empowers international standards is that their development process, as well as the certification procedure followed to assess conformity to a defined standard and the overall institutional framework providing trust via accreditation to certifiers, replicates to some extent a principle of delegation of territorial sovereign rights. As we will see in Chapter 4, the claim to abide by formal mandates is particularly important in the functioning of international standardisation bodies such as the Comité européen de normalisation (CEN) and, to a lesser extent, the ISO. It also often rests on formal oversight of state agencies on tests and conformity assessment procedures. Conversely, the exogenous logic is related to the diffusion of rules through market mechanisms first. In this case, what empowers an international standard is its use by market actors backed by inviolable contracts across sovereign spaces. The fact that the standard is set by an official standardisation body with a defined mandate or, on the contrary, by a private organisation or even a consortium of large multinational enterprises setting their own technical specifications for the larger use of the market is irrelevant, so long as users are ready to comply with a standard guaranteed by contractual market relations. As Chapter 7 on service offshoring in India will make clear, this is why, for instance, management officials in charge of quality may see no difference of status between ISO standards and management methodologies devised by American specialised firms, such as the CMMI Institute.

In brief, compliance to standards within the confines of a hybrid space rests on the ambiguity of a mixture of endogenous and exogenous logics. Whereas the contradictory logic of sovereignty has always been split between territorial power and transnational legal guarantees in support of capitalist expansion, the deterritorialisation of sovereignty in transnational hybrid authority marks a shift away from the endogenous pole of sovereign power towards more exogenous means of supporting the current accumulation regime.

* * *

This chapter built on the concept of transnational hybrid authority to explain how and why standards reflect a non-conventional form of power that puts the existing structure of the interstate system to the test. This first supposed answers to the question of why the power of standards is so often referred to as a shift towards hybrid forms of markets and regulation. A growing literature in law and political science has emphasised that this reflects the increasing complexity of regulatory environments made of a comprehensive web of public and private actors in charge of setting technical specifications and assessing that goods and services fully comply with them. More generally, I stressed that this reflects a core ambiguity in the notion of governance with regard to the exercise of authority: a way to exercise power over a defined domain and population without, however, the plain attributes of sovereign rights. In both cases, the notion of hybridity is first and foremost used as a *default attribute* that says more about what it is not than what such non-conventional forms of power are. In contrast, in answering the question 'what are hybrids?", I emphasised that ambiguity ought to be considered a *substantial – or ontological – attribute* of the non-conventional forms of power and regulation embodied by standards in the context of globalisation. In contrast to conventional understandings of standards as science- and expert-based instruments devised to respond to an increasing complexity of the world or a distinguishing feature of private market power and capture, ambiguity appears as a defining criterion of standards in conferring authority to new actors on a number of new issues across sovereign spaces in the context of globalisation. As a mythological creature of sorts, or a post-colonial subject, a transnational hybrid authority gains credence despite its undefinable imaginary with multifaceted and contradictory dimensions, for each of its parts is made of real and well-defined features along the three dimensions analysed in this chapter. Finally, answering the question 'how hybrids work', the final section of the chapter delineated an analytical framework for the study of the transnational hybrid authority of standards. The power of standards should thus be viewed from a tri-dimensional perspective. First, it confers authority across an institutional continuum that blurs the distinction between private and public actors. The scope of this authority, in turn, extends along the material continuum of physical measures and societal values in such a way as it undermines the divide between humans and non-humans. Finally, the recognition of this authority lies on conflicting sources along a spatial continuum, where the endogenous logic of territorial sovereignty becomes interwoven with an exogenous logic that reinforces the deterritorialisation of capitalism.

In the conclusion of this book I shall return to the broader implications of this analytical framework. We shall see in particular why such ambiguity should not be considered only as a challenge to sovereignty, to democratic representation, and to the interstate system supporting the expansion of market power in the contemporary restructuring of the capitalist regime of accumulation. Following a Coxian understanding of critical approaches to international political economy (Cox, 1981, 2002), my analysis remains situated in time and place and should facilitate an assessment of the current potential for emancipatory transformation and change. As ambiguity supports a form of authority used to govern, but at the same time can be brought into play to engage and resist forces supporting such power, it should thus also be viewed as providing opportunities for those struggling for progressive change. In contrast to the dematerialised idealistic reading of some post-colonial scholarship and, to a lesser extent, social studies of science and technology, Best, for instance, sees ambiguity rather as a mundane everyday practice that can become a means of resistance (Best, 2008, 2012: 87, 2013). All in all, standards put in motion new informal institutions that intermingle with the interstate system in many different ways to form an ambiguous juxtaposition of power instances. The rise of such transnational hybrid authority is, like all hybrids, essentially Janus-faced. Yet, far from being resistance discourses and strategies studied by post-colonial scholarship or symmetric quasi-objects flattening the relation between science, society, and nature, the transnational hybrid authority of standards remains very much anchored to the material asymmetry of power that underlies the global governance project of unifying markets across sovereign territories.

3 Service Offshoring

The New Frontier of Globalisation

The death of outsourcing that KPMG, one of the 'big four' consulting firms, announced a few years ago did not go unnoticed in the small world of business process outsourcing (BPO), where firms rationalise operations by taking advantage of lower wage costs in countries such as India. According to KPMG, 'there is a revolution taking shape in the business services industry, one that disregards the traditional shared services and outsourcing paradigms' (Justice, 2012).[1] While the report of the death of outsourcing might be exaggerated, it does suggest that companies are becoming more aware of the difficulties of depending on work done on the other side of the world in a context marked by shrinking labour cost arbitrages and growing concern for quality, security, and intellectual property rights. Whatever the truth might be, this stands in stark contrast to the view shared at the turn of the millennium and the heyday of overstatements on globalisation. At that time, there were endless accounts of the glorious hopes for the twenty-first-century entry into the world of service offshoring, with no shortage of relocation of activities abroad. If the sky was not the limit, labour cost differentials, skills, digitisation, and the ability to codify well-defined segments of tasks performed at distance were. As Alan Blinder, former Vice-Chairman of the Fed, put it: 'The old assumption that if you cannot put it in a box, you cannot trade it is hopelessly obsolete. Because packets of digitized information play the role that boxes used to play, many more services are now tradable and many more will surely become so. In the future, and to

[1] To avoid confusion between outsourcing and offshoring, it is worth remembering here that outsourcing describes the purchasing of goods and services from outside specialist providers at arm's length either nationally or internationally. In contrast, offshoring describes purchases of goods and services from foreign providers at arm's length or the transfer of particular tasks within the firm to a foreign location, i.e. to foreign affiliates. The cross-border aspect is the distinguishing feature of offshoring, i.e. whether goods and services are sourced abroad as opposed to the domestic economy, not whether they are sourced from within the same firm or from external suppliers (the aforementioned definitions are drawn from OECD, 2010: 220).

a great extent already, the key distinction will no longer be between things that can be put in a box and things that cannot. Rather, it will be between services that can be delivered electronically and those that cannot' (Blinder, 2006).

These opposing views are not only about the extent to which the expansion of the tertiary sector is likely to follow the rise of a global knowledge-based economy – or put differently: how a greater internationalisation of services would better match the large share of services in GDP and employment. They also reflect rising concerns on the tradability of service activities, concerns which typically focus on employment and barriers to trade. For instance, while Blinder minimises many overstatements on the expected overall loss of service jobs in the United States, he does stress that services likely to remain in the United States would for the most part be low-skilled with clear negative effects on wages (Blinder, 2006: 124). Others, on the contrary, suggest that higher-skilled and higher-waged service jobs in the United States are a comparative advantage and provide an enormous opportunity to make the case for service trade liberalisation, especially in removing non-tariff measures in India, China, and the European Union (Jensen, 2011). In the face of such difficulties in exporting services, service internationalisation is often considered to rely less on trade than on foreign investment (Enderwick, 2007). While not immune to the burden of domestic regulatory environments, foreign investment is viewed as a way to overcome restrictive border measures as long as 'national treatment' rules out discrimination between foreign and domestic firms. This drove the early moves of companies such as American Express, Accenture, and IBM to India in the 1990s. It is also why Indian service firms such as Tata Consulting Services and Infosys have now for some time established affiliates in the United States and elsewhere.

The role of standards in the internationalisation of services attracts far less attention than job losses, trade wars, and the competition of new multinational services firms. And yet, in codifying the disaggregation of service production and delivery into discrete and independent stages – likely to be assessed against distinct quality performances and security guarantees – standards have a direct effect on employment, trade, and investment. A number of accounts consider standardisation of services as a crucial requirement in developing a competitive advantage through the substitution of capital for labour and by establishing routine labour processes suitable to less skilled and cheaper employees abroad (Zeithaml et al., 1990: 79; Johnson and Nilsson, 2003; McIvor, 2010). Some might identify this as service innovation, while others emphasise the engrained labour alienation that such practices imply.

Similarly, there is a common understanding that trade in services is quite different from goods and relies on standards (for quality, safety, protection of consumers, etc.) often embedded in domestic regulation and likely to impede market access (Boden and Miles, 2000b; Djellal and Gallouj, 2010; World Trade Organization, 2012; Du Tertre, 2013). From this point of view, standards often protect public interest and fulfil policy objectives; they can also be used as tools for market integration, in particular for services close to manufacturing like those provided in huge and remote back offices.

Against this background, what is the relationship between the expansion of the tertiary sector, the globalisation of production and market networks, and the authority conferred on standards in the regulation of contemporary capitalism? More specifically, what is the role of non-conventional forms of regulation such as standards in service offshoring? Conventional explanations focus on information asymmetries, institutional factors, and the sectorial specificity of trade transactions in services as compared to goods. As we will see in further detail in the third section of this chapter, they distinguish, in particular, between intangible and relational services resistant to standardisation and so-called industrialised services likely to be standardised in complement to relying on information and communication technologies. This is, in other words, a *restrictive hypothesis*, according to which the attributes of service and domestic institutions will largely determine the propensity for standardisation and internationalisation. This chapter aims to consider the wider potential power plays in shaping the political economy of standards as they may encourage or hinder offshoring. In contrast to such sectorial- and institutional-dependent views, I propose an *extensive hypothesis*. I argue that standards can accommodate opposing political economy objectives and power configurations. By linking the global marketplace to distinct national economies, service standards can respond in diverse ways to quality and security guarantees. They can incorporate specialised knowledge into packages of segmented tasks and apply worldwide market discipline to unskilled workers. But there is no reason to believe that they cannot also lead to a more progressive understanding, for instance by providing safety procedures to workers heading to the night shift in remote customer centres, or protecting consumers with guarantees, or procedures for the handling of complaints and dispute resolution (such as with the ISO standards 10002 and 10003). Such an ambiguous form of non-conventional power in the regulation of the international economy reflects what I call a transnational hybrid authority.

My extensive hypothesis allows for appraising the socio-political implications of standards in the internationalisation of services along the three analytical dimensions discussed in the previous chapter. The range of actors and standardisation bodies having authority to set international standards is vast; depending on their instruments, they span the institutional continuum of the public and private spheres. Moreover, the scope of objects likely to be standardised spreads across a material continuum, whose two poles are the physical and societal worlds. In other words, even if service offshoring depends on standardised technical interfaces supporting the provision of highly industrialised services, it cannot ignore shared social and cultural values. Finally, the system of standards recognition overlaps a wide spatial continuum; following the dual nature of sovereignty, the recognition of standards is both endogenous to the territorial State and exogenous, akin to the market forces of transnational capitalism. From this point of view, standards per se neither support nor hinder the internationalisation of services. It all depends on which actor exerts authority to set such and such a standard likely or not to be recognised across such and such jurisdiction. It involves actors with opposing political economy interests and values, concerns issues intrinsically blurring the frontier between societal and physical worlds, and rests on a system of recognition that reinforces the deterritorialisation of sovereignty. In a nutshell, setting standards and complying with them is inherently ambiguous.

This chapter begins by sketching the contextual and conceptual background of services and the expansion of the tertiary sector, emphasising the distinct regulatory constraints required of services. I suggest that issues of quality and security, conventionally seen as the heart of the regulation of services, should be understood as social institutions, whose qualification remains highly political. The second section examines in some detail the available data on the internationalisation of services. It unpacks what I refer to as the 75/25 puzzle: services account for 75 per cent of GDP and employment in rich countries and are considered the key to future development of a global knowledge-based economy; meanwhile, over the last two decades, their overall share in the global economy has continued to represent around 25 per cent of world trade and investment. It also sheds light on a significant shift in composition (with developing and emerging countries having doubled their share), with growing diversification up the value chain and a profound integration of services into manufacturing processes. The third section focuses on conventional explanations of drivers and barriers of greater services tradability, with a distinct emphasis on standards within the broader regulatory environment of contemporary capitalism. It discusses in particular the restrictive hypotheses on the internationalisation of services put forward by institutional economics and approaches inspired by the

French *régulation* theory. Finally, the chapter examines the extent to which services may defy the very idea of standardisation and support a more extensive hypothesis. To this end, service standards are examined as particular instances of transnational hybrid authority whose ambiguous power spreads across sectorial and institutional specificities. They define a wide range of quality and security requirements likely to have strong social and political implications.

The Test of Tertiarisation

The extension of the tertiary sector is one of the most striking aspects of the shift in recent decades towards a so-called knowledge-based global economy. Services now account for approximately 75 per cent of GDP and employment in the advanced economies of the OECD, and more than 50 per cent in developing countries and emerging economies. As we have seen, services raise contested issues pertaining to employment, trade, and foreign investment. While they have exhibited the strongest growth in the global economy for many years, increasing doubts are emerging around the continuity and sustainability of this movement. This leads us to consider with renewed caution assertions made over the last twenty-five years on the importance of this phenomenon and the transformation it implies for the transnational regulation of global capitalism. Karpik (1989) associated the shift towards services as part of a new 'economy of quality and singularities'. In the same vein, Castells (2001: 56) identified the advent of a 'service society' as a radical shift towards 'informational capitalism'. And according to Gadrey (2010), an 'economy of quality, service and knowledge' calls for the construction of a whole new 'economy of care'. The rise of services directed towards the immediate and tailor-made satisfaction of consumer needs is, the argument goes, most certainly going to give rise to 'new forms of competition' (Petit, 2008). Finally, some see the valorisation of knowledge in the service economy as marking nothing other than a new accumulation regime: 'cognitive capitalism' (Moulier-Boutang, 2007).

The first step needed to stand back from the spin generated around the 'new service economy' is to look at what the terms 'services' and 'tertiary sector' really mean. What are services? This is not a trivial question. Even before initial negotiations attempting to establish a regulatory framework within which to liberalise trade in services had begun in the mid-1980s, the renowned British weekly *The Economist* had come up with the following oft-cited formula: services are 'things which can be bought and sold but which you cannot drop on your feet' (*The Economist*, 1985). Such a definition does not, of course, do full justice to a plethora of activities difficult to pin down on this basis. Still, it catches in a nutshell what bodies responsible for producing and harmonising international statistics have taken years to define in order to reach a comprehensive approach to statistical issues entailing services, guiding negotiations, and supporting implementation of international agreements encompassing services such as the General Agreement on Trade in Services (GATS). This was in particular the task of the new United Nations Manual on Statistics of International Trade in Services, first published in 2010. The manual restates the usual cautionary note on the heterogeneous range of service activities difficult to encapsulate within a simple definition and remains all the more true given the difficulty in separating services from the goods with which they may be associated in varying degrees.[2] That being said, it defines services as follows: 'the result of a production activity that changes the conditions of the consuming units, or facilitates the exchange of products or financial assets' (United Nations, 2010: 8).

A prime interest of this apparently simple definition is to extend the meaning of international trade in services to encompass various modes of supply included in the GATS, in particular the supply of services through the commercial presence of affiliates established abroad. More substantially, it allows for flexibility in responding to opposing conceptions of service activities. Most frequently cited characteristics of services are intangibility (or immateriality such as in teaching or health), co-production (or the relational dimension between a customer defining its needs and a consultancy firm offering a response), heterogeneity (or the idiosyncratic dimension of services ever customised according to clients' needs and thus supposedly avoiding standardisation), and perishability (that is, the impossibility to stock services in an inventory like goods) (Millar and Choi, 2011: 28). The UN definition clearly abandons the criteria of immateriality and intangibility, which for two centuries had stood at the heart of classical and neoclassical economics. Instead of seeing services as what Adam Smith and his followers saw as residual, non-productive, and immaterial activities, characteristic of non-productive labour, the definition echoes the diversity of service activities and their increasing integration with the production of physical goods. It is from this perspective that it draws on Hill's seminal contribution (Hill, 1977). In contrast to neoclassical analyses, Hill put the focus on the

[2] The definition used by the manual is based on the 2008 version of the System of National Accounts jointly published by the United Nations, the International Monetary Fund, the World Bank, the Organisation for Economic Co-operation and Development, and the European Commission.

processes involved in the transformation of the state of an individual or object and the relational issues involved when the activity implies co-production between the producer and the consumer of the service. From this focus on changes in the condition of a person or an object, the impossibility of stocking services has nothing to do with a physical impossibility supposedly derived from immateriality; it is, rather, a 'logical impossibility[, since] a stock of changes is a contradiction in terms' (Hill, 1977: 319). It follows that, as services are exchanged without transfer of ownership in contrast to goods, 'models of pure exchange economics of a Walrasian type in which existing goods are traded between economic units are quite inapplicable and irrelevant to services' (Hill, 1977: 318). This is why a medical service is not some kind of immaterial drug, a training course some kind of immaterial brain, or a life insurance policy some kind of immaterial grief. Hill's definition has been widely recognised as the most suitable for both research and statistical purposes. It allows us to stick to as simple a definition as possible for our own purpose. Suffice here to mention a last point, made by Gadrey (2003: 18) in order to account for a greater variety of demand rationales characterising services. According to this view, services are deeply embedded in social institutions. They span two axes. The first concerns the 'type of capacities' with which the user mainly interacts. It opposes technical capacities (such as the automatic teller machine providing you with cash) to human capacities (such as those displayed by a discrete Swiss banker advising you in a cosy meeting room on how to park cash and dodge tax authorities). The second axe is about the 'mode of request' chosen by the user to get a service delivered. It opposes interactive requests (i.e. to 'be served' by a service provider such as a customer-relations officer based in a call centre in Bangalore) to non-interactive requests (i.e. to 'serve yourself', when you use an IT platform, pay for a nice trip on your own, or do your own cleaning).

This said, there is little disagreement about which activities account for services or not, but much more on how to establish an explicit and comprehensive definition (Illeris, 2007: 24). We accept Bryson and Daniels' idea that 'it is important not to become too distracted by the search for a precise definition of services' (Bryson and Daniels, 2007a: 4). Even more important is the extent to which a service economy is in fact inextricably linked to the manufacturing industry. In his classic work, Daniel Bell (1973) saw the dawning of the post-industrial era as an age in which services would replace manufactured goods as a growth engine. Consolidation of service activities in economies the world over has not, in fact, led to a major change in traditional industrial production. The situation is not that of a closed system in which an increasing share of service activities results in a corresponding decline in manufacturing or agricultural production. On the contrary, production systems are increasingly interdependent between manufacturing industry and services. Two obvious examples present themselves. The smartphone, which has made Apple the world's most valuable company and the first-ever company to reach a value of $1,000 billion on the US stock market, is as much a product manufactured by the millions with cheap labour in Foxconn's Chinese factories as a product of branding and design services praising its trendy and friendly way to be connected to the latest 'killer app' in town. For far longer, lift companies have based their profits on a similar overlap between manufacturing and services: much of their added value comes not from the production of lifts, but from service contracts, especially when these are mandatory, as is the case in most developed economies. More generally, services play a core support role in integrated production networks, which rely on transport and logistics, and complex financial and insurance products. The symbiosis of all business services – themselves heavily reliant on information and communication technologies supporting knowledge creation, innovation, and timely processing and dissemination of information – with manufactured goods has become a preponderant reality.

The major change in the structure of our economies lies in the fact that service activities now constitute a fundamental, but not exclusive, dimension of economic activity. The debate on the 'tertiarisation of industry' and the 'industrialisation of the tertiary sector' (Rubalcaba, 2007) has given way to the notion of 'servicification' or 'servitization', where services and manufactured goods should be approached as mutually interdependent, as they are increasingly traded as a package (Kommerskollegium, 2012; Cernat and Kutlina-Dimitrova, 2014). With high-value added services such as design, R&D, and architectural and engineering services concerned, there is clear evidence of the difficulty involved in dealing with a phenomenon which involves both interdependency and integration. Just as the production of manufactured goods can no longer be envisaged independently of all the supporting services upstream and downstream the production chain, industrial and commercial logic is contributing to the increased industrialisation of services. Before Apple's smartphones even existed, Boden and Miles pointed out: 'The service economy is not merely an economy in which service sectors are quantitatively dominant. It is one where "service" is becoming a guiding principle throughout the economy' (Boden and Miles, 2000a: 258). This is undoubtedly why Bryson and Daniels allude to 'a complex process of hybridisation whereby the categories of manufacturing and services are becoming increasingly blurred' (Daniels and Bryson, 2002; Bryson and Daniels, 2007a: 7). A decade later, ISO's presentation of its strategy for services explicitly refers to the '"hybridization" of companies' offering … [as a reason] to look more closely at evolving market needs in the service sector' (International Organization for Standardization, 2016b: 8).

The integration of service activities into the economy as a whole assumes that this can be achieved in parallel with the upsurge of the international division of labour over the last two decades or so. In a context characterised by a globalisation of production networks and a common use of information and communication technologies (ICTs), the internationalisation of services has thus become a major issue. Traditionally, the phenomenon of internationalisation was confined to physical goods. Now, with sufficient infrastructure, huge quantities of data, information, and digitised documentation can be exchanged instantly, at minimal cost, anywhere in the world. As we will see in further detail in the following section, the development of this infrastructure has profoundly changed the tradability of services across borders whilst imparting major significance to the notion of knowledge within the service economy. Delivering services and coordinating knowledge on a worldwide scale also assumes increasing movements of factors of production, including the circulation of individuals, financial flows, and direct investments in cross-border trade in services. This has now taken on a new dimension with the massive recourse to big data, cloud computing, the Internet of things, and platforms which use hardware and software resources to deliver all sorts of services. To a certain extent, the Holy Grail would be to take the internationalisation of services full circle; it would abolish the factors of time and space hitherto seen at the heart of a service relation, which in turn often justifies a distinct regulatory environment usually viewed as hindering internationalisation.

However revolutionary the technological shift of the cloud is and notwithstanding the localisation and materialisation of data-farms, the global and timely delivery of services still depends on an architecture of legislation, regulations, and standards which plays a determining role, mandatory as well as voluntary, in establishing the conditions of access to the market. The instruments devised to regulate the internationalisation of services are highly diverse, ranging from intergovernmental organisations, and supervisory and standards bodies to bilateral and multilateral framework agreements, as well as ad hoc stakeholder conferences and roundtables, informal institutional platforms, and industrial and/or non-profit consortia. I shall examine this in full detail later in the chapter. For the time being, suffice it here to emphasise that in one way or another, all of them run up against the difficulty raised previously: how to establish commonly accepted criteria to specify the expected characteristics of a service and define a benchmark against which conformity to a promised service may be judged. Despite an undoubtedly fragmented environment across nations and industries, the issue generally remains the same: define the quality of the service.

In classical and neoclassical market theory, quality is not disputed by the agents, who are held to have the same representation of the item being traded – a representation founded on the supposedly complete information provided by the price signal. Scholars of asymmetries of information have developed a fresh view on the notion of quality as an independent and determining variable in the markets, distinct from self-regulatory markets based on price information. In what has become a classic article of economics, Akerlof (1970) showed the fundamental information asymmetry characterising a market using the example of used cars, in which the seller has information about the goods which the buyer does not possess. Such asymmetry prevents the market for physical goods working properly; in addition to 'brand names' and 'state licences', 'guarantees' on product quality are identified as 'counteracting institutions' likely to help rectify this (Akerlof, 1970: 549–550).[3] George Akerlof, Michael Spence, and Joseph Stiglitz would later receive a joint Bank of Sweden Prize in Economic Sciences in Memory of Alfred Nobel for this demonstration. They show the extremely narrow condition of validity of the fundamental hypothesis of neoclassical Walrasian economics, only valid if all goods and services exchanged have a homogenous quality, perfectly defined and of common knowledge to all agents. A further hypothesis makes inroads into uncertainty about quality, shared by all agents, particularly with regard to products whose past or – more importantly in a context characterised by the emergence of new markets – future is unknown (Hirschman, 1970; Lupton, 2005).[4] Some management research analyses this hypothesis in greater depth by drawing a distinction between quality based on personalisation and that based on industrialisation. In the first case, the emphasis is on customer satisfaction, while the second addresses issues such as the reliability of processes of production and service provision (Deming, 1981; Sundbo, 2002). Working from the idea that a service-based relationship involves co-production between the provider and the beneficiary, the analysis focuses on the additional uncertainty resulting from the co-incidence of these two types of quality (Parasuraman et al., 1985; Grönroos, 1990;

[3] For further discussion, see: (Stiglitz, 1987; Orléan, 2011: 87).

[4] More generally, the question of how uncertainty affects markets goes back to the classical distinction that Frank H. Knight (1971: chap. 7) made between 'true uncertainty' and 'risk', subject to be numerically measured and anticipated with objective data.

Johnson and Nilsson, 2003). The salient point here is the degree to which quality is always a bone of contention, whose ambiguous status lends itself to be tweaked in various ways and subject to controversy.

French *régulation* theory has provided several analytical tools for these questions. One of them is that quality uncertainty calls for a specific mode of regulation, distinct from price. While Fordism put price and volume at the core of the mass production and consumption of uniform standard goods, the economic focus now considers quality as a prime form of competition. In order to respond to this puzzle, quality can be conceived as an institution in the sense of the institutional economics of John Commons (Allaire and Lemeilleur, 2014). Cautious as ever on the 'uncertainty of meaning of the word institution', Commons' definition is as simple as it is far-reaching: an institution is 'Collective Action in Control of Individual Action' (Commons, 1934: 69). In sharp contrast to neoclassical economics focused on rational individuals isolated in a state of nature, the individual with whom Commons is dealing is thus an 'Institutionalized Mind' (Commons, 1934: 73). The quality of a good or a service therefore can neither derive from a price signal nor from any intrinsic attribute of such good or service. On the contrary, it should be viewed as a social construct stemming from power relations between private and public actors who pursue their ever-evolving interests. This view of quality as an institution with explicit and codified procedures at the crossroads of power and interests calls for conventions fixing implicit anticipations and coordination expectations. This is clearly more than just providing information, which can later be passed on to the consumer via a nice label. It stems from complex negotiations, through which a series of institutional forms, regulation agreements, conventions, and standards constitute and situate the qualitative attributes of a given product or service. The actors involved in this process struggle to impose a concept of quality that, following Commons' definition, allows for collective action to control the individual action of agents involved in economic transactions. This prompts Chanteau (2011: §24) to describe standardisation in the field of quality as a 'total social fact, in which exchanges of goods and signs, as well as ways to control individual behaviours play out in such a way as economic and political facts become indivisible'.

This discussion of quality is not specifically focused on services, though. In contrast, Callon et al. (2002) put forward a semantic shift from quality to qualification that opens up major implications for analysing service activities. While *régulation* scholars draw on Commons to emphasise the power relations involved in the institution of quality implemented by third-party certification or participatory guarantee systems, the concept of qualification describes power issues following the idea of product differentiation first developed by Chamberlin (1925) in his *Theory of Monopolistic Competition*. Chamberlin emphasised the growing importance of the ability of sellers to differentiate products from one another; furthermore, he considered that the ability to modify the list of qualities of a product should be seen as a strategic resource for firms. According to Callon et al., this process of qualification/requalification is at the heart of the practices involved in the service industries. As service activities are viewed as the basis of the economy of qualities, they depend on differentiated modes of regulation. Markets for services are often based on the expectation that consumers actively participate in the re/qualification of products. Their regulation should accordingly counterbalance the power position given to the seller by the possibility of monitoring users, observing their preferences, and targeting the products offered to them. According to the authors, nowhere is this biased collaboration between supply and demand more obvious than on the Internet. Whilst the authors wrote at the beginning of ICT-supported singularisation of products, Internet-supported services are now part of our daily life. Data ownership and privacy protection have become prominent issues for all sorts of applications used on smartphones, such as Google map, Uber, Booking.com, Amazon, and the like – all adapting their offer to your own, and unlike any other consumer's history. This new mediation of the consumer in the trust and judgement-based tools of services has clearly demystified the notion of 'quality of service' and made it possible to envisage changes in the many and various mechanisms used to judge, evaluate, sanction, and grant trust. The cognitive resources of the users are mobilised in such a way as to engage debates previously monopolised by scientists in economics and, to some extent, other social sciences. As Callon et al. (2002: 96–97) note, qualifying products in a service economy is likely to provoke 'a profound transformation of the rules by which markets function … The organisation of markets becomes a collective issue and the economy becomes (again) political'. As qualification practices are assumed to generate disputes of all kinds as to the nature and level of trust required in the market transactions of services, they regard 'hybrid fora' as promising avenues for debates on the organisation of such markets. As we saw in the previous chapter, hybrid fora are arena that mix expert and lay knowledge for debating technical choices involving the wider public (Callon et al., 2001); here the technology is less about 'quasi-objects' that bracket nature with culture, than the qualification devices designed to build trust between service providers and consumers. While this might be conceivable at the local level of communities, it is much harder to imagine at the global level that service offshoring has reached today. Before discussing the relationship among qualification, standards, and regulation, this prompts us to examine in more detail the issues at stake with the internationalisation of services.

The 75/25 Puzzle

In the introduction to this chapter we saw how forecasts of the overall number of jobs threatened by the ability to relocate services abroad caused widespread alarm in the early years of the 2000s. More than a decade has passed since the heyday of globalisation narratives that fed such expectations. In the meantime, the economy has been struck by a global crisis whose extent was only rivalled in the 1930s. Job losses for the less skilled and, increasingly, the middle classes in rich countries have remained a core issue. While part of the onus falls on manufacturing, services are seen as bearing the brunt of the expense. Big firms led the way, using the Internet to offshore information technology, back-office work, and even much more complex tasks to places such as India, Morocco, the Philippines, and Fiji. Smaller firms soon followed suit. The scope of service activities identified as 'tradable' across borders has widened as well. *The Economist* (2013), widely known for its liberal views (in the UK sense), recognises that 'offshoring has brought huge economic benefits, but at a heavy political price'. At the same time, services are still seen as remaining inherently difficult to internationalise. Companies have begun to reconsider what many saw as common sense. In contrast to offshoring – a relocation of activities abroad, mainly for cost purposes – the new game in town, in particular in the United States, is reshoring – a relocation of activities at home, not only for quality purposes but also with increasingly competitive costs. Against this background, let us take a closer look at what is at stake, first by examining existing data on the internationalisation of services.

Services now account for approximately 75 per cent of GDP and employment in the advanced economies of the OECD, and more than 50 per cent in developing countries and emerging economies. Nevertheless, we are still far from a globally integrated supply of services. Sectorial coverage remains narrow and apparently no upsurge of total trade in services has taken place in the last two decades. As Tables 3.1 and 3.2 show, services continue to represent less than 25 per cent of world trade. What has happened is a significant 'shift of composition' between developed and developing countries. Over twenty years, developing countries have almost doubled their share to reach around 30 per cent since 2012. This share of services in world trade, under 25 per cent, remains in stark contrast to the 75 per cent of GDP and employment in

Table 3.1 *Exports of goods and services, 1990–2017 (US$ at current prices and current exchange rates, in millions)*

	1990	2000	2010	2017
Total trade in services	831,345	1,521,978	3,896,263	5,351,191
Total trade in goods	3,429,185	6,418,666	15,077,595	17,408,800
Share of services in %	19.5	19.2	20.5	23.5

Source: UNCTAD, UNCTADstat2018 (series discontinued for 2017, with a compilation based on the IMF Balance of Payments and International Investment Position Manual, Sixth Edition (BPM6); yet this has minor impact on figures shown on the table).

Table 3.2 *Share (%) of developing/developed/transition economies of services exports, 1990–2017 (US$ at current prices and current exchange rates)*

	1990	2000	2010	2017
Developed economies	79.9	75.5	69.2	68.2
Developing economies	18.1	23.1	28.5	29.5
Transition economies		1.3	2.3	2.3

Source: UNCTAD, UNCTADstat2018 (series discontinued for 2017, with a compilation based on the IMF Balance of Payments and International Investment Position Manual, Sixth Edition (BPM6); yet this has minor impact on figures shown on the table).

the advanced economies of the OECD. This is what I call the 75/25 puzzle of the internationalisation of services.

Only a portion of those activities involves international trade transactions *per se*, with tasks outsourced according to an arm's length principle. A significant number take place within the network of multinational firms, between parent companies and their affiliates, described in the jargon as 'captives'. This prompts us to pay additional attention to foreign direct investments (FDI). At first sight, the overall share of services in FDI has not changed so much either. Figures presented with some caveats by UNCTAD a few years ago (Table 3.3) show that their share increased by less than 10 per cent over twenty years, of which trading and finance account for more than half. However, here again it is worth noting an important shift in composition: while developing countries accounted for less than 20 per cent of all FDI inward flows in services in 1990–1992, that figure reaches more than 40 per cent twenty years later, with the share of business services having almost doubled. Consulting, accounting, auditing, data management, customer relation

Table 3.3 *Estimated world inward FDI flows, by sector and industry, 1990–1992 and 2009–2011 (US$ millions)*

	1990–1992				2009–2011			
Sector/industry	Developed countries	Developing economies	Transition economies	World	Developed countries	Developing economies	Transition economies	World
Total	134,419	39,779	1,530	175,728	729,143	613,772	82,593	1,425,507
Primary	10,215	4,211	911	15,337	43,994	75,884	14,733	134,611
Manufacturing	37,422	14,457	279	52,158	161,241	155,722	14,528	331,491
Services	77,605	17,918	208	95,732	475,660	369,913	52,830	898,403
of which: Trade	16,735	2,474	22	19,232	61,126	51,463	13,803	126,392
Finance	25,745	2,575	15	28,335	194,735	77,595	9,322	281,652
Business activities	17,107	4,257	130	21,494	154,803	149,066[a]	18,029	321,898[a]
Share of services (%)	57.7	45.0	13.6	54.5	65.2	60.3	64.0	63.0
of which in trade and finance (%)	54.7	28.2	17.9	49.7	53.8	34.9	43.8	45.4
of which business activities (%)	22.0	23.8	62.4	22.5	32.5	40.3	34.1	35.8

Notes: a. A considerable share of investment in business activities is in Hong Kong (China), which accounted for 37 per cent of developing economies and 17 per cent of the world total during 2009–2011. Hong Kong (China) data include investment holding companies.

Data should be interpreted with caution. The world total was extrapolated on the basis of data covering 79 countries in 1990–1992 and 116 countries in 2009–2011, or the latest three-year period average available. They account for 83 and 90 per cent of world inward FDI flows, respectively, in the periods 1990–1992 and 2009–2011.

Source: Adapted from UNCTAD, *World Investment Report 2013.*

Table 3.4 *Domestic services value added share (%) of gross exports (2011)*

United States	United Kingdom	France	EU27	Japan	India	China	Hong Kong, China	Rest of the World
49.8	52.1	51.0	42.7	45.0	47.9	27.7	76.4	24.5

Source: OECD-WTO Trade in Value Added (TiVA) – December 2016.

centres, all belong to these new types of business services easily established in developing countries and attracting massive volumes of foreign direct investments. Interestingly, sectors such as health or education, despite often making headlines, remain marginal in comparison, with worldwide inflows of $391 and $814 million, respectively, in 2009–2011. Certainly data on services, let alone data on their international trade, are notoriously complex to come by. The figures are thus likely to remain disputable, as data classification and accessibility remain elusive.

A joint OECD/WTO initiative has attempted to address this issue by producing data disaggregated by the value added in the exchange of goods and services consumed worldwide. According to the methodological note accompanying the publication of the statistics, 'the break-up of domestic content by direct and indirect sectoral value added reveals that a large chunk of the value originates indirectly from service sectors' (OECD-WTO, 2013: 11). Well aware that international trade remains dominated by goods even while a large share of global GDP and employment accrues to services, the new metrics produced by the OECD/WTO initiative clearly shed additional light on the significance of services (see Table 3.4). While these estimates are dated, they show that the service sector contributes around 50 per cent of total exports from countries such as the United States, the United Kingdom, and France, with the whole European Union, Japan, and India scoring in the same average, and even close to one-third in the case of China, often seen at the bottom end of low-cost manufacturing value chains. The metrics also provide some estimates of the service content of overall manufactured goods (cf. the aforementioned discussion regarding 'servitization'). Here too, the figure is significant, with an average of one-third in the reference year of 2009, corresponding to an increase of between 5 and 10 per cent in many countries since 1995.[5]

[5] OECD-WTO: Statistics on Trade in Value Added, online at: https://stats.oecd.org/Index.aspx?DataSetCode=TIVA_2016_C1, last accessed 10 August 2018.

Against this background, it is quite obvious that the offshoring of services is a powerful and significant phenomenon of contemporary capitalism. The shift began in the 1980s with outsourcing contracts in data processing and call centres at the bottom of the value chain. Today, it has moved into much more advanced sectors, with activities such as legal, fiscal, or medical services, financial consulting, and all sorts of business services enabled by information technology, from the entertainment industry to security-related activities. Although the development was still embryonic then, UNCTAD emphasised its significance more than ten years ago: 'the cutting edge of the global shift in production activity [gives] rise to a new international division of labour in the production of services' (UNCTAD, 2004: xxv). While firms have embarked on offshoring in many areas, at the same time they have become increasingly aware of the difficulties to be overcome.

The evidence established so far lends us to consider that the jury is still out concerning the rise of service offshoring. In contrast to largely inflated estimates made in the heyday of expanding globalisation, the 25/75 puzzle has not completely disappeared from the picture. The ratio of trade and foreign investment to overall employment and GDP figures in services has remained relatively stable over the last quarter century. There is, however, a significant shift in composition, as developing and emergent countries have more than doubled their share in trade and inward foreign investment – even if the recent development of re-shoring may suggest that this share has reached a tipping point in some domains. Moreover, there is ample evidence of greater diversification, as the offshoring of services has shown a propensity to climb the value chain, especially in sophisticated business services. Finally, compiling a thorough inventory of the internationalisation of services remains plagued with difficulties. At the same time, the integration of services deep into manufacturing instils a significant service-content of exported manufactured goods. How shall we explain, then, the expectations and on-going difficulties with the expansion of services across borders? Moreover, to what extent do these developments prompt new expectations on the role of standards in contemporary capitalism? This is what the following sections examine.

Restrictive versus Extensive Hypotheses

Many services were long considered non-tradable since they required buyers and sellers to be in the same place at the same time. A radiologist was thus supposed to be located in the hospital where the patient had her/his x-ray done. Similarly, an insurer relied on the knowledge of local agents to assess risks of local firms. ICT has removed many of these constraints and made services much more tradable. So-called digitability is the foremost driver of service offshoring. Not only can all kinds of information be digitally stored and directly exchanged almost anywhere in the world, but computer technology also allows knowledge to be digitised, codified, fragmented, and re-organised in such a way that the production process can be spread across the globe in complex, disparate, and far-away locations.

While ICT is both a basic infrastructure and a sophisticated knowledge-intensive tool of service offshoring, the second chief driver is the quest for cheap labour costs. Management textbooks make that point time and again (McIvor, 2010; Oshri et al., 2015). We have seen how controversial this can be in terms of jobs losses and flexibility for middle-class workers in industrialised countries, as well as low-paid and exploitative jobs in developing and emerging economies. Conversely, the narrowing of labour cost differentials between industrialised countries and developing and emerging economies accounts for a large part of the 'reshoring' of services to the United States (*The Economist*, 2013). Labour costs thus remain a core dimension in the arbitrage of going or not going offshore. Implications in terms of wage erosion, shifts in bargaining power detrimental to labour, and redeployment difficulties towards higher value-added jobs have not, by a long way, disappeared.[6] The ability of service providers and consumers to move is a third factor to take into account. Some services rely heavily on the mobility of experts or basic workers (what the GATS defines as 'mode 4 – movement of natural persons'). This occurs in the many instances when foreign nationals provide services abroad either as independent suppliers (for instance, accountants) or as employees of a firm (for instance, a construction company). Similarly, services such as education, tourism, and health treatments rely on the mobility of consumers ('mode 2' of an international exchange of services, according to the GATS definition). Moreover, the mobility of service providers depends in many instance on legal and regulatory provisions that set out the conditions for affiliates and subsidiaries to establish a commercial presence ('mode 3' of the GATS, relating to foreign direct investment and other forms of foreign-owned and controlled companies). Finally, many studies stress the importance of looking beyond ICT, labour costs, and the mobility of service providers and consumers in order to take due account of language and cognitive skills, cultural understanding, and various kinds

[6] See Levy (2005) for further discussion on this critical issue.

of geographical links that, though harder to quantify, have increasingly been recognised as important supports of the tradability of services. As Bryson and Daniels emphasise in reference to Peter Dicken's bestselling study on globalisation, 'unlike the first "global shift", the geography of the second shift is determined by the education and language abilities of services workers located in low-cost location' (Bryson and Daniels, 2007a: 12). This is why, for instance, a large multinational insurer would have a greater propensity to develop micro-insurance policies in a country with some shared knowledge and interpretation of what a supply or demand for cover really means to protect low-income people. And this is probably what promoters of micro-insurance have in mind when they emphasise that 'slow and steady incremental improvements are … important for fostering a culture of insurance in low-income markets and creating a firm foundation for future expansion' (Churchill and McCord, 2012).

Despite these drivers, many analyses stress the factors that continue to hinder the internationalisation of services. A first explanation considers the sectorial specificity of services, whose intrinsic characteristics are seen as an insurmountable obstacle to internationalisation. A particular instance in this regard is the fact, discussed earlier, that some service activities cannot be stored and require direct co-production between clients and suppliers. Similarly, the more services tend to be immaterial, the argument goes, the harder it is to provide them at a distance. In addition, most firms providing services are SMEs and thus more likely to face additional difficulties in projecting their activity internationally. Often, to complement the sectorial account, another explanation focuses on the institutional specificity of services. In this perspective, as the intangibility of many services industries carries with it the risk of market failures and behaviour taking advantage of market power, governments have a greater tendency to intervene in the regulation of markets for services than for goods. Economists usually describe this as regulation driven by political economy considerations. They also take into account the public policy objectives involved in government interventions targeting services directly or indirectly related to the public interest. Such regulatory practices aim at environmental issues, consumer protection, health and safety guarantees, the provision of basic universal services, securing professional skills such as those of doctors, lawyers, accountants, and many others. The distinct legal framework and the pervasiveness of the regulatory environment surrounding a wide range of services are thus seen as 'non-tariff measures' which can, if not properly checked, prompt major hindrances to internationalisation (Copeland and Mattoo, 2008). In a report closely examining such non-tariff measures and their embeddedness in trade policies, the WTO goes straight to the point: 'Given the pervasiveness of services regulation and its commingling with trade protection a clear identification of which measures are trade restrictions, or a neat separation of the protective component in such measures, is fraught with difficulty' (World Trade Organization, 2012: 78). Ironically, this has not prevented the OECD from launching in 2014 a project aimed at scoring and weighting barriers to services trade in twenty-two sectors across forty-four countries so as to build a new Services Trade Restrictiveness Index (STRI) and identify what the organisation sees as 'potential scope to unlock growth through regulatory reform' (OECD, 2017).

The details of what is driving or hindering services offshoring is not only becoming steadily more sophisticated but also highly contestable. Beyond sectorial explanations focused on the specificity of service industries and institutional explanations examining the range of instruments available to governments to pursue policies that can or cannot be trade restrictive, such non-tariff measures often rely on standards, testing, certification, and labelling in order to claim scientific rather than political justification.[7] Conventional explanations of drivers and barriers to services offshoring often fail to stress the extent to which the expansion of the tertiary sector has prompted new expectations on such non-conventional forms of power and regulation in contemporary capitalism. In the light of this, it is important to understand the influence of mechanisms that go well beyond intergovernmental cooperation and trade transactions. Indeed, greater global integration in the supply of services hinges upon a number of informal, non-state processes challenging national regulatory arrangements. It is in this context that international voluntary standards come into play.

From a number of angles, economic analysis has studied the conditions in which standardisation is possible and the resulting consequences for the pursuit of growth and innovation in the service sector. From a microeconomic and rationalistic point of view, standards are instruments designed to improve the reliability of the market for services, by certifying that providers supply services to users according to the agreed terms (Zeithaml et al., 1990; Johnson and Nilsson, 2003). According to Blind, it is precisely 'because of the intangible nature of services and the

[7] According to the compilation of reported non-tariff measures from ninety-nine countries established by the Office of Economics of the United States International Trade Commission (USITC), standards, testing, certification, and labelling rank second among specific measures (after government procurements) if one excludes generic measures such as import-, export-, investment-related measures setting out local preferences, restrictions, or prohibitions (Eaton et al., 2013: table 2).

information asymmetries thus caused between management and service provider, [that] the need to introduce quality standards for each stage of the service production is especially high' (Blind, 2004: 167). Where possible, solutions incorporate enhanced use of technology, in particular ICT (Barras, 1986). Where it is not, tasks can still be disaggregated into codified and standardised methodologies. From this managerial perspective, standards describe the 'extent to which tasks in a process can be executed using a set of consistent and repeatable steps' (McIvor, 2010: 105). Service standardisation is viewed as completing codification, which in turn provides a complete description of tasks along distinct components. A flurry of analyses has examined the scope of services likely to be standardised according to various taxonomies specifying such sectorial determinants. A number of them link the technological characteristics of services to the nature of the market they serve to define patterns of firms' choices between standardisation and customisation (Boden and Miles, 2000b; Tether et al., 2001; Djellal and Gallouj, 2002; Djellal and Gallouj, 2010). Moreover, while studies generally suggest that the information intensity of a service is correlated to its relational intensity and thus potentially hinders its prospective disaggregation, standardisation, and internationalisation, some scholars see it the other way round: they argue that high information intensity makes an occupation more amenable to disaggregation since each process can be well defined and thus codified and standardised (Mithas and Whitaker, 2007). In either case, while focussing on strategic interactions within a market environment in order to determine innovation and competition patterns, microeconomic studies of service standardisation overlook the political economy content which make these processes fraught with contradictions and power practices beyond and across firms' behaviours and particular industries.

Scholarship inspired by the French régulation theory systematically includes social and political issues which can promote internationalisation of services in its analysis. Following on from Baumol's pioneering work (1967), Petit has analysed how the extension of business services at the interface of economic and political spheres contributes to a new growth regime which is imbalanced, dualistic, and based on a form of competition predominantly influenced by changes in consumer habits and lifestyle. Accordingly, the development of standards is hindered by the lack of shared values to orient and appreciate the qualitative changes required for the expansion of the tertiary sector. For instance, venture capitalists remain relatively weakly internationalised, as they suffer from the absence of a certification system that could help them standardise procedures for pooling investors and entrepreneurs (Petit, 2007: 95–96,

2013). As seen earlier, Gadrey's typology also focuses on the variety of demand rationales in services. While his account deals primarily with the industrialisation of services, rather than their internationalisation, it also implicitly assumes strong institutional determinants. According to this analysis, an institutional environment with social and gender inequality is likely to support a commodification of services that seek to maximise productivity, economies of scale, and standardisation processes. The richest social group concentrating a large share of national revenue would have a clear interest in being served with industrialised and certified processes by a large (and mainly female) reserve army of poor workers. In contrast, greater social and gender equality is likely to lead to a more progressive understanding of service rationalisation supporting more reflexivity in working routines, and a rejection of commercial services and technologies that promote labour and gender inequality. The axiom would be 'serve yourself' rather than 'be served' in a standardised and commodified way, bringing with it important limits to market access on an international scale (Gadrey, 2003: 105ff). For his part, Du Tertre (2008, 2013) examines the institutional outcomes that result from the distinct labour relations in services, in particular immaterial and relational activities (such as training, consultancy, and medical services). Here again, uncertainty as to quality and usefulness is understood as the most direct hindrance to the establishment of standards likely to support their internationalisation. According to Du Tertre, the responses to this uncertainty highlight two opposing types of institutional outcomes. On the one hand, there is the option of neo-Taylorist standardisation, designed to reduce the time needed for establishing the relationship between provider and beneficiary. This development, based on industrial logic, favours the use of machines (for instance, automated teller machines instead of bureaux de change) and information and communication technology (e-banking), as well as the formalisation of stereotypical behaviour (the number of refusals a catalogue retailer should accept before halting communications). The opposite approach is that of professionalisation, including the formulation of ethical principles which commit providers and beneficiaries alike. While this is common practice in the national framework of regulated professions such as in engineering or health services, Du Tertre sees it as largely non-existent when it comes to the international offshoring of services. Irrespective of the favoured institutional outcome, his argument goes, production of a service should always go hand in hand with a 'social relation of accessibility', defined as a 'historic and institutional construct' characterised by considerations such as geographical proximity, temporal synchronisation, and cultural and social understanding (Du Tertre,

2008: 70–71). In this perspective, the internationalisation of services is rather unlikely, unless the trading logic remains close to that for manufactured goods (Du Tertre, 2013). In a similar vein, Mouhoud et al. have established a typology of services, based directly on their links with territories (Mouhoud et al., 2010).

According to institutionalist scholarship inspired by the French *régulation* theory, the uncertainty inherent in the intangible and relational nature of many service activities should not be apprehended as a problem of information asymmetry skewing the price mechanism, but as the logical consequence of the actual conditions in which wage relations and forms of competition are implemented in a post-Fordist regime of accumulation. Uncertainty as to quality and usefulness reflects the very heart of a service characterised by a high relational and immaterial component. From this perspective, it is quite normal that uncertainty should hinder the establishment of standards promoting the internationalisation of such activities. Standards would have a role only for services close to manufacturing such as those provided in the huge, remote back offices which Du Tertre calls 'information factories' (Du Tertre, 2013: 116, n. 8). This industrialisation-based reasoning favours the use of machines along with information and communication technologies, and assumes the imposition of stereotyped behaviours. All other types of services, however, will run up against socialisation processes – the social relation of accessibility as Du Tertre would call it – reflecting the disparate nature of institutional and localised dynamics. Here, the service-based logic specific to the immaterial and relational nature of the activities in question acts as a hindrance to the development of standards likely to support internationalisation.

Thus, French *régulation* scholarship sheds light on the socio-political underpinning of potential standards supporting the internationalisation of services. In doing so, it posits a restrictive hypothesis that paradoxically loses sight of political economy power plays. Such a restrictive hypothesis proposes that standardisation and internationalisation are closely defined by the attributes of the service and domestic institutions. In this perspective, standardisation and internationalisation of services indeed rest, on the one hand, on sector specificity, according to which the more the service provided corresponds to the ideal type of a relational, non-material service oriented towards the end consumer, relying on high-intensity labour, the less likely it is to be standardised and internationalised. On the other hand, it also follows institutional specificity, according to which the closer a service is to this ideal type, the harder it will be to find a substitute in the national territorial framework within which service relations are institutionalised, as is the case for regulated professions in medicine and law. This analysis offers a relevant framework for explaining the structural reasons underlying the difficulty of establishing internationally recognised standards in order to facilitate the offshoring of the most relational and intangible services. However, in my view, its hypothesis regarding the conditions for standardisation and internationalisation of service activities is too restrictive because it makes them dependent on sectorial and institutional specificity that rejects broader power configurations.

In sectorial terms, the specificity of 'productive configurations' and forms of competition for activities close to the ideal type of relational, non-material services oriented towards end consumers and relying on high-intensity labour does not in principle exclude rationalisation as a substitute for face-to-face transactions and objectivation of the high cognitive component of such transactions. Like money and law in traditional Marxist analysis, quality standards could represent a new general equivalent in a capitalist economy (Hartmann, 2013). There is no better example than the battery of international quality and security standards for management systems developed by ISO from the 1980s onwards. These are still the best-selling standards in the world, and thus arguably the most widely used.[8] Mention may also be made of management instruments of similar nature marketed by many private players, such as Motorola's Six Sigma method (which aims to identify and drastically reduce standard deviation from all the given specifications in a production process). Rationalisation of a service relation may thus apply to many highly relational services such as in education, healthcare, and consulting, for which many standards exist and support a market increasingly outsourced on an international scale.[9] As I shall explore in further detail in Chapter 7, it is by following this logic in particular that India has become the world's office, providing offshore services on an ever-higher rung of the value chains.

[8] Cf. the ISO 9000 family for quality in general, the ISO 14000 family for management of the environment, the ISO 26000 family for social responsibility, the ISO 27000 family for information security management, the ISO 31000 family for risk management, and most recently, the ISO 51000 family for energy management.

[9] For example, in 2014, Panasonic's robotic bed that turns into a wheelchair, cleared an international safety standard for care robots, reducing liability risks for the company (see: 'Robots: the ghost in the machine gets smarter', *Financial Times*, 7 December 2014). For their part, requirements for personal financial planners are set by the ISO standard 22222:2005; with regard to education, there are many private certification concerning ICT skills and the ISO also published in 2010 a new international standard (ISO 29990:2010) for providers of non-formal education organised outside the framework of the formal education system, for activities such as training courses offered by a company or an association.

On the institutional level, a restrictive hypothesis on standardisation and internationalisation of services arises from a highly local and territorial conception of institutional dynamics. Territorial and local roots of the socio-historical construct of economic activity, and the weight of political action engaged in at this level, certainly have an influence on the expected development of services, even in the field of ICT conventionally viewed as the least territory-based.[10] This also applies to national environments and institutional complementarities as highlighted in studies inspired by the French *régulation* theory (Aoki, 1994; Amable, 2000). However, there is no reason why these considerations should be exclusive. A service relation may *also* be supported by procedures that are formally documented, certified, and accredited beyond the regulatory scope or political action of local, regional, and national authorities. This could include – and thus question – measures relating to social or gender inequalities and the power plays these suppose. For instance, it is precisely with this in mind that a quality standard for customer contact centres in France (NF 345) has been devised, some parts of which have been used at the European level (EN 15838:2009). Analysis of the controversies surrounding the establishment of ISO standard 26000 on social responsibility may be conducted from a similar perspective (Ruwet, 2009; Capron et al., 2010; Barraud de Lagerie, 2011). In both cases, the institutional base of the service relation rests on a polymorphous and partially transnational space. The 'social relation of accessibility' of a service may come into play at this scale.

It is against this background that I propose an extensive hypothesis in order to overcome sectorial- and institutional-dependent explanations of services standardisation likely to support internationalisation. Viewing the nature of the service and its embeddedness in national institutions as determining factors does not fully do justice to the great variety of responses that international standardisation is likely to provide to the intangible and relational dimensions of many types of services. Service standards can link national economies to the global marketplace by responding to quality and security uncertainties in many, let alone opposing, ways. Far from stereotypes that deny cultural and labour issues involved in the service economy, the prospects of services offshoring rests on state and non-state regulatory arrangements which are not necessarily isomorphic. My extensive hypothesis emphasises a range of quality and security requirements likely to be standardised notwithstanding sectorial and local institutionalised specificities. Service standards can thus

[10] This phenomenon is known to geographers as the paradox of the digital economy (Morgan, 2004; Brette and Moriset, 2009).

accommodate opposing political economy objectives and power configurations. On the one hand, they can promote a broadening and deepening of minimal market rules; on the other, they can include a number of provisions with the aim of defining labour-, environment-, gender, and any other socially-based specifications likely to be instrumental to more vulnerable actors in the production process and to ensure differentiated usage of services. Such an ambiguous juxtaposition of power instances confers authority to new actors on all sorts of new issues across sovereign spaces. This prompts us to explore the extent to which standards are a key instrument of transnational hybrid authority likely to have strong social and political implications in the quest for regulatory convergence supporting market access to services.

Standards-Defying Services?

In the introductory chapter of this book I defined standards as voluntary technical specifications explicitly documented and published as tools for the organisation of production and exchange of goods and services. Standards thus codify technical specifications regarding measurement, design, performances, as well as side effects of products, industrial processes, and services. I also emphasised the considerable overlap between mandatory standards embedded in regulations set by public authorities and voluntary specifications set by standard-setting bodies not based on sovereign State authority such as the International Organization for Standardization (ISO). 'Standards hover between state and the market, to quote Schepel once more ... standards are very rarely either wholly public or wholly private, and can be both intensely local and irreducibly global ... standards can be seen as links between these spheres and institutions' (Schepel, 2005: 4). The relationship between such ambiguous juxtapositions of standard-setting agencies and society are thus controversial, in particular with regard to services, where deep cultural and societal values and elusive labour issues are very likely to be involved. What is more, deregulation, liberalisation, and privatisation are likely to prompt new service standards to compete with previous rules governing public utilities.[11]

One of the clearest ways in which standards contribute to a non-conventional form of power is in their ability to reinforce path-dependant oligopolistic trajectories in technological innovation (David, 1985; Mock, 2005). Such a political economy of innovation inspired by

[11] The introduction of the book provides further detail on existing scholarship on standardisation.

Schumpeter has led Dudouet et al. (2006) to conceive standardisation as a process of appropriation with significant incidence on market power and competition. Short of exclusive exploitation rights included in patents, standards codify technical specifications that can de facto exclude alternative technology and contending processes. Innovative technology conforming to such standards often includes patented technology, for instance in large-scale and forward-looking IT-enabled business services relying on extensive databases and complex algorithms. The neo-Schumpeterian analytical lens provides a persuasive explanation why large firms are likely to use standards as artefacts for maintaining their domination on distinct technologies. It is less focused, however, on how the power of such standards redefines the articulation between economic and political spheres. Accordingly, the following question remains unanswered: how do institutions underpin the power of such standards or, in other words, which institutional fora are privileged to exercise such power? Moreover, the approach tends to overcome the extent to which standards are used as levers of power on technological innovation, and reproduction is likely to vary according to distinct objects and processes. This leads to a second unanswered question: how does the issue to be standardised affect the power configurations at stake? For instance, to what extent can the establishment of sustainability standards – including labour, environment, and gender provisions – be clearly distinguished from standards established as instruments to develop technical interfaces to industrialise the provision of services? Finally, neo-Schumpeterian approaches do not explicitly examine how international standards may undermine the constitutive principles of the territorial sovereignty of states. Conversely, the role that states and intergovernmental organisations play in supporting the authority of international standards deserves further attention. This leads me to a third question: on which transnational space can technical specifications be defined, diffused, and recognised among sovereign states and, subsequently, what transfer of authority does this imply for contemporary modern liberal democracies?

These three questions on who standardises what and where bring me back to the analytical framework set out in the previous chapter. The ambiguous juxtaposition of power instances set in motion by what I call a transnational hybrid authority overcomes conventional oppositions. As Busch point out, 'private standards and public regulations are two similar and sometimes overlapping forms of governance' (Busch, 2011: 27). Following the weight of common belief in the separation between the public and private spheres in modern capitalism and liberal democracies, legal provisions still often contradict this understanding and confine most international standards to the domain of private voluntary market tools provided by private firms.[12] It is worth noting in this regard that the WTO analysis of non-tariff measures has largely gone beyond this narrow understanding. The following definition from its 2012 Report is unequivocal: 'Although cast as "voluntary" in nature (because they are imposed by private entities), private standards (i.e. ISO & other standards) may become *de facto* a necessary condition for market access even if not imposed by law. ... the effect of a particular private standard, if pervasive, could be greater than that of a government regulation of a smaller country' (World Trade Organization, 2012: 211). Moreover, the ambiguous power of standards rests on the multiplicity of arenas in this domain and their wide range of institutional status (Djelic and den Hond, 2014). The multiple standards authorities entail numerous agents who play or claim to play a role as new actors gain power to regulate a wide range of issues recognised as such across borders.

Consequently, the nature and implications of standards shaping the internationalisation of services can be appraised along the three-dimensional framework of analysis exposed in the previous chapter. This prompts us to aggregate the three following categories (Figure 3.1): the actors – i.e. who has the authority to set standards; the objects – i.e. what is standardised; and the space – i.e. where and whence standards are implemented. As we have seen, expectations regarding quality, security, and capacity to deliver services relate both to the private sphere of economic activities governed by market constraints and the public sphere of political action in the general interest of society. At the same time, they are closely intertwined with societal values and face regulatory diversity across countries. These dimensions of the transnational hybrid authority of service standards have some resemblance to what Busch has called the tripartite standards regime.[13] They provide a comprehensive yet differentiated picture of the wide range of actors involved in setting standards, the breadth of issues concerned, and the deterritorialisation of sovereignty conveyed by the multiple systems of certification and accreditation worldwide. We will examine those three dimensions in more detail hereafter.

The first dimension of the framework is the *institutional continuum* of actors in the public and private spheres. As we have seen, there is much ambiguity on situating standards along this continuum and that ambiguity plays its part in conferring authority on a range of actors previously denied

[12] For instance, The Swiss Federal Act on Technical Barriers to Trade (946.51 – art. 11) limits State support or direct participation in standards-setting bodies to standards referred in public law and therefore acquiring a status of mandatory state regulation.

[13] The regime includes standard-setting, accreditation, and certification (Busch, 2011). Those three processes "traverse and integrate public and private spheres both within and across nations" (Loconto and Busch, 2010: 508).

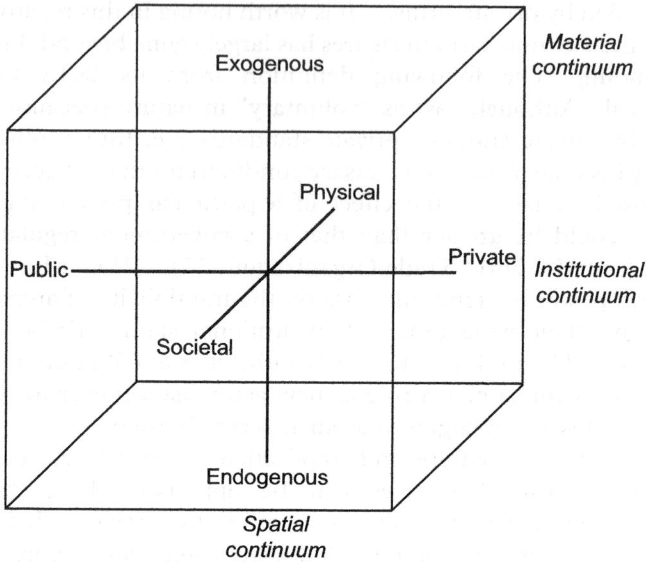

Figure 3.1 Transnational hybrid authority

such capacity. Market mechanisms and policy choices both affect the agents involved in standardisation, but in various ways. Technical specifications belong to the private sphere of economic activities governed by market constraints; they affect social and technological change from that angle. However, they remain related to the public sphere of political action directed toward the general interest of society – for instance, by determining a certain level of risk and setting principles of liability or, *a contrario*, by allowing rent-seeking behaviour and market power. Hence, even in the circumscribed field of technical specification, standards relate as much to capital accumulation and technical progress as to social improvement or various instruments of the welfare state. When they are mandatory, enforceable, and general, technical specifications become part of public law and enjoy the status of government regulation. While some environmental, health and safety performances are defined under such procedures, they are often established by non-governmental actors on a voluntary basis. In such cases, technical specifications involve standard-setting bodies, whose private or public statutes vary considerably from one country to another. The wide range of actors and standardisation bodies able to set international standards are likely to address the distinctive aspects of the service sector in various ways. This is precisely the issue in understanding what can be standardised.

The second dimension charts the *material continuum* that outlines what can be standardised along the two poles of the physical and societal worlds. In aggregating the relationship between human beings and nature, technical specifications range from natural and invariable physical measures to constructed and historically bound societal values. Following an approach instigated by the Actor Network Theory in social studies of science and technology, Busch also views standards across the modern divide among nature, science, and society: 'standards permit us to create complex socio-technical networks. As people are used, people and things are tested, and we shall determine what shall count. Those people and things that pass the tests or make the grade are drawn into various networks' (Busch, 2011: 12). This dimension highlights the extended scope of international standardisation. Long thought of only for physical specification such as those set for nuts and bolts, standards now cover all sorts of issues with clear and present implications for the societal world. One example among many is the European Technical Specification CEN/TS 16880:2015 for service excellence that sets out guidance 'in order to create outstanding customer experiences, exceed customer expectations and achieve customer delight'.[14] As regards services more generally, this indisputably raises further pressing questions about what is a service standard. Do service standards concern the material support enabling the delivery of services (protective equipment used in the leisure sector, IT interface of call centres, etc.)? Do they concern mere procedural and generic aspects of services irrespective of the cultural context of their provision (e.g. billing, complaint redress, information provision, security requirements in the domain of business processes continuity)? Are they able to take into account more substantial expectations related to cultural and societal values, as well as labour processes including implicit skills embedded in the co-production of intangible and relational services (special needs for disabled people, the elderly, and children, or customer-oriented behaviour specifications for employees in tourism)? Clearly, the internationalisation of the service sector depends on standards for the development of technical interfaces to industrialise the provision of services. Yet it also relies on shared cultural values involved in the relationship of co-production between producers and consumers. Service standards can follow various paths in addressing such relationships, whether on a sectorial basis, taking basic cultural and societal values into account, or on a generic and horizontal basis, reflecting stereotyped behaviour that denies cultural and elusive labour issues involved in co-production activities. While those specifications can be defined as voluntary or

[14] See CEN Project Committee 420 on Service Excellence Systems.

mandatory on a national plane, if included in international standards they must be recognised beyond state borders.

The third axis of our analytical framework is the *spatial continuum* where the jurisdictions that support the system of recognition of standards overlap. While international standardisation is driven by attempts to homogenise technical specifications across national jurisdictions in order to reach a higher level of market and regulatory convergence, it faces a plurality of standards or tools for assessing conformity with them. From this point of view, compliance to standards is ambiguous. It rests on the dual nature of sovereignty: the endogenous logic of the territorial State and the exogenous logic of transnational capitalism. The endogenous principle depends on a system of obedience in agreement with the territorial space of state sovereignty. It supports a system of compliance to standards established on the assumption that a development process and an assessment of conformity procedure based on territorial sovereignty is what confers authority to an international standard. This is why, for instance, only one standard-setting body per country is eligible to membership in the ISO, even if many of them are non-governmental bodies identified as the 'most representative of standardisation in their country'. When it comes to official international standard organisations, we are clearly faced with an ambiguous mixture of private and public bodies. ISO procedures make it even more ambiguous, as the actual work of defining standards is done in working groups where private experts sit in their own right, in contrast to plenary meetings of technical committees, composed only of national delegations. At the other end of the spectrum, when it comes to the exogenous principle guiding the compliance to standards, it is first and foremost their use by market actors across the globe that defines their spatial diffusion. A flurry of standards are set and used away from any defined system of national representation and delegation, and nevertheless entrenched in sovereign contract law. This is particularly the case for so-called consortium standards that define an agreed specification, usually openly accessible, yet developed by a restricted number of market players. They first dealt primarily with specifications required for the development of new products (like CD-ROM or DVD), or with interoperability in the field of ICT (such as the TCP-IP protocol used for connecting computers into a single worldwide network). Recently, however, a range of initiatives appear to promote such standards with a much broader scope. Corporate social responsibility benchmarks and sustainability labels in global value chains are cases in point (such as the Roundtable for Sustainable Palm Oil (RSPO) and the Roundtable for Responsible Soy (RTRS)). A number of initiatives also affect the service economy more directly, such as the Global Reporting Initiative (GRI) guidelines on presentation of information, specifically designed to make the environmental and social impact of major companies more visible. As a final point, it should be noted that, in the absence of harmonised regulation or standards, mutual recognition of standards and regulations reflects a step toward the exogenous end of the continuum. Mutual recognition is in fact designed to ensure that governments recognise each other's standards and accept results of testing, inspection, certification, or accreditation bodies in specified industries. In Europe, for instance, products not subject to technical harmonisation at the EU level should in principle be exchangeable freely so long as they conform to domestic specifications and regulations of a member state. As we will see in the next chapter, this is also one of the key issues in provisions on regulatory convergence and non-tariff measures in the new generation of mega-trade deals.

* * *

This chapter has shown that the growing significance of services in the world economy involves being able to reach some common understanding on the quality and security of service provision. Far more than a simple question of asymmetry of information, the qualification of service activities is intrinsically political and relies on broader social institutions. It involves new patterns and agents of change through formal and informal regulatory practices of a wide range of non-state actors. In this context, service standards are likely to play a crucial role. Most existing literature shares, however, a restrictive view on the potential for standardisation of services, which would narrowly depend on domestic institutions and intrinsic characteristics of the industry concerned. This book takes a broader view. It proposes an extensive hypothesis that goes beyond a sectorial- and institutional-dependent analysis. The question is not whether services standardisation can or cannot occur across sectors and various domestic institutions. It is that when it does, it rests on the ambiguity of the statutes of actors setting such standards, of the issues eventually standardised, and of the space on which they are recognised. Service standards thus reflect a form of transnational hybrid authority that blurs the distinction between private and public actors, whose scope extends from physical measures to societal values, and that reinforces the deterritorialisation of regulatory practices in contemporary capitalism. That is the analytical framework within which the following chapters provide an overview of developments in the field of service standardisation. I start with the institutions which provide authority to standards as *de jure* or *de facto* regulatory instruments supporting the expansion of the tertiary sector across borders.

4 Standards as Regulation

Standards are often heralded as supporting innovation and leading to greater rationality and coherence in distinct industries, services, and organisations. Yet all of them give rise to on-going struggles in complex configurations of power involving multiple actors, including multinational corporations, organised interests, and state regulators. In other words, standards benefit from massive transfers of authority to bodies situated between the political and the economic spheres, serving as alternatives to conventional state regulation. How, then, do distinct institutional environments affect the development of standards likely to support the internationalisation of services? And regarding the supposed specificity of services, how distinct is the authority of standards in the domain of services compared to goods?

We saw in the previous chapter that many services are generally described as intrinsically resisting relocation because of their intangibility and their involvement in activities supposing a co-production between producers and customers. Service standards would, therefore, be considered a sticky case of standardisation. However, a shared assumption is that, although international standards in the service sector appeared only very recently, they are expected to surge in parallel with the importance of services in the economy and society at large, with growing reliance on standards in a context of regulatory convergence, races towards innovation, and a more intense internationalisation of the sector (Blind, 2003; ISO, 2016; Vries and Wiegmann, 2017). As emphasised some time ago by a former Secretary General of the ISO, 'one of [the] biggest challenges is precisely how to address the service sector'.[1] In any case, the growing importance of service standards tests existing differences between, on the one hand, the ISO and the European environment and, on the other hand, the American institutional framework for setting standards. The former favours a coordinated standardisation system with greater

[1] Interview with Alan Bryden, Secretary General of the ISO (2003–2008), Geneva, 8 June 2007.

reliance on territorially based legitimacy and state oversight; the latter gives preference to competing sources of standards and relies on market mechanisms to ensure their definition and adoption. Analyses present the two systems as a case of 'regulatory competition' (Czaya and Hesser, 2001; Tate, 2001; Werle, 2001; Mattli and Büthe, 2003; Winn, 2009). More broadly, it should be underlined from the outset that the contentious new generation of preferential trade agreements such as the aborted Transatlantic Trade and Investment Partnership (TTIP) between the United Sates and the European Union include regulatory convergence, if not harmonisation of standards, among their core objectives. From this standpoint, future developments of service standards are more likely to depend on divergent national institutional frameworks than service sector specificity.

Yet, this does not help us to uncover the power relations underpinning various forms of standards supporting deeper integration of the market for services. The three-dimensional analytical framework set out in the previous chapters serves this purpose, by bringing together more systematically the plurality of actors defining the standards, the distinctiveness of services concerned by standardisation, and the transnational space in which such standards are likely to be recognised and implemented. From this standpoint, service standards embody a transnational hybrid authority that confers on them increasing power to regulate contemporary capitalism. They blur the distinction between private and public actors in charge of setting rules; their scope spans a continuum bringing together physical measures and societal values; and they reinforce the deterritorialisation of regulatory practices in contemporary capitalism. In contrast to conventional views opposing the American system to the ISO/European framework, the chapter argues that the ambiguous juxtaposition of power instances set in motion by the most recent institutional developments of service standards is likely to face trade-offs and compromises reflecting contrasting models of standardisation, not only between, but also across, those systems. While this undermines the conventional analysis of a transatlantic divide in standardisation, it also shows that the variance between product and service standards is much greater in the European context and the ISO system than in the United States, where it is hardly debated.

This chapter looks at the various institutions providing authority to standards as de jure or de facto regulatory instruments governing the internationalisation of services. It is arranged as follows. The first section provides background on the institutional environment of standardisation and introduces the case of the transatlantic divide. Sections 2–4 present, respectively, the ISO setting, the European, and the

American systems. A subsequent section reviews how recent negotiations on mega-trade agreements reinforce the essential role of standards in further market integration. A final section discusses the evidence provided in the chapter more specifically in relation to the three institutional, material, and spatial continuums on which the power of standards rests.

The Institutional Environment

The previous chapter discussed at length how the dominance of services can be seen as one of the most striking aspects of changes in the world economy over recent decades. Today, services account for around 75 per cent of all jobs and GDP in OECD economies – and over 50 per cent in developing countries and emerging economies. While total trade in services has remained constant for the last two decades, developing countries have almost doubled their share in the world trade in services to reach more than 30 per cent in recent years. The significance of services goes beyond their growing share in the economy and close connection to technology and knowledge. It is also intimately related to an expected surge in their internationalisation resulting from durable regulatory reforms. An institutional environment enabling the internationalisation of services has gradually emerged with the application of the General Agreement on Trade in Services (GATS) in 1995, negotiations underway at the World Trade Organization (WTO), and the adoption in 2006 of a new EU directive (2006/123/EC) on services in the internal market. Moreover, as we will see in further detail in this chapter, preferential trade agreements, including the new – and highly controversial – generation of mega-trade deals such as the Canada-European Union Comprehensive Economic and Trade Agreement (CETA), the aborted Transatlantic Trade and Investment Partnership (TTIP) between the United States and the European Union or the Comprehensive and Progressive Trans-Pacific Partnership (CPTPP) among Asia-Pacific countries, specifically target convergence in regulatory approaches, harmonisation of standards, and growth of investment and trade in services as crucial issues.

In the past, technical specifications were largely the preserve of the regulatory framework of law, company standards set by managers, and, to a marginal degree, national standards institutions. Today, the regulatory framework of law has yielded ground to voluntary standards drafted by a raft of international or regional public and private sector bodies. The creation of the WTO in 1995 was a crucial threshold. Unlike the GATT, whose provisions in terms of technical regulations were not very restrictive, the Agreement on Technical Barriers to Trade (TBT), the Agreement on Government Procurement (GPA), the review of the Agreement on Sanitary and Phytosanitary Measures (SPS), and the General Agreement on Trade in Services (GATS) grant international standards a major role in the harmonisation of technical specifications applicable to goods and services. State regulation in this domain must comply with 'legitimate objectives'. With regard to goods, such concerns are related to health, safety, and environmental issues. In contrast, as we have seen, conflicting understandings of market uncertainties about quality and security are the major issues in the sphere of services; they encompass a wide range of expectations regarding, in particular, competence and professional skills, the capacity to deliver business continuity, data protection and privacy, and consumer protection and information, as well as larger societal and environmental concerns. As the WTO is not a standard-setting body, its promotion of regulatory convergence is made by prompting its members to use international standards. GATS article VI:4 thus assigns to the Council for Trade and Services (through its Working Party on Domestic Regulation) the largely market-inspired task of developing 'any necessary discipline' to ensure that regulation by states is not 'more burdensome than necessary to ensure the quality of the services'. Article VI: 5b specifies that in this respect, 'account shall be taken of international standards of relevant international organisations'. According to the WTO, regulatory cooperation in services would have much to gain from improving 'regulators' understanding of, and confidence in, standards and requirements with which they may not be familiar' (World Trade Organization, 2012: 186). Similarly, in the wake of earlier guidance, the OECD published in 2012 a new Recommendation on regulatory policy and governance suggesting that members, 'In developing regulatory measures, give consideration to all relevant international standards and frameworks for co-operation in the same field and, where appropriate, their likely effects on parties outside the jurisdiction'(OECD, 2012, recommendation # 12). Yet, existing provisions still grant a wide range of international bodies the ability to define on their own terms standards affecting the internationalisation of services.

In the United States, standardisation is usually presented as fragmented and organised on a sectoral basis. A variety of competing standards organisations (formal and informal) set market-driven standards exempt from state intervention. The system follows a so-called model of direct participation, where companies have direct access to standard-setting activities with international claims. In contrast, the European standardisation system is coordinated and centralised, and operates under a higher degree of government control. The European standardisation bodies[2] follow a so-called model of national participation, where a national body holds the voting rights within umbrella standardisation bodies such as the CEN (except for the United States, the system is similar for non-European standardisation bodies members of the ISO). In spite of their differences, the European and American standardisation systems have common characteristics. Both rely on private organisations to shape standards on a voluntary basis. They follow a due process open to all interested parties and their deliberations are based on the 'state of the art'. The draft standards are subject to public consultation and the general interest is supposed to prevail over particular interests. Finally, their standard-setting bodies recognise the primacy of international standards, even though the understanding of what 'international' means remains controversial. Despite these similarities, several conflicts remain between ISO/European and American standards developing organisations (SDOs).

From the American point of view, the national participation model in the European standardisation bodies gives them a substantial advantage at the international level (Zuckerman, 1996: 40; Czaya and Hesser, 2001: 32). The Vienna and Dresden agreements between the ISO and CEN, respectively with the International Electrotechnical Commission (IEC) and the European Committee for Electrotechnical Standardization (CENELEC), can indeed be seen as benefiting European actors, as they grant provisions for a simultaneous recognition of standards at the European and international levels (with CEN potentially leading the work) and have ensured a coordination of the standardisation work between those organisations. Moreover, with about 4,000 European standards indirectly referenced through 30 directives, the New Approach allows for a presumption of conformity with essential requirements for all firms that claim to be using such standards; but clearly, there will be more European than American firms doing so (ASTM International, n.d.)!

In reverse, from a European point of view, the decentralised and fragmented standard-setting procedures in the United States represent a barrier to the US market. Moreover, American SDOs' claims to serve the public interest often hinder strong commercial interests and

[2] The three European standardisation bodies are: the Comité européen de normalisation (CEN), the Comité Européen de Normalisation Electrotechnique (CENELEC), and the European Telecommunications Standards Institute (ETSI). The ETSI differs significantly from the CEN and CENELEC in that it accepts corporate as well as national members. For further analysis of the European context, see: (Egan, 2001; Schoechle, 2009: 24).

contending regulatory competition. Finally, the international reach of standards developed in the United States tends to undermine the authority of formal standardisation arenas such as the ISO and CEN.

Unsurprisingly, scholars have discussed such transatlantic divergences on the most appropriate institutional foundation of international standards at great length (Schmidt and Werle, 1998; Abbott and Snidal, 2001; Czaya and Hesser, 2001; Egan, 2001; Nicolaïdis and Egan, 2001; Spruyt, 2001; Tate, 2001; Werle, 2001; Mattli and Büthe, 2003; Vogel, 2009; Winn, 2009: 21; Mattli and Büthe, 2011). Yet Egyedi questions such a clear-cut transatlantic divide in standardisation (Egyedi, 2005). She stresses that this tends to underestimate the opening of most industry consortia and overestimate the democratic institutional pledge of formal organisations. While committees in both cases are formally open and work on a consensus-oriented basis, stakeholders with few resources, whether in civil society organisations or small and medium-sized enterprises, continue to take pains to participate in standard-setting practices undertaken in technical committees. Thus, it is important to overcome the conventional caricatures opposing the American and ISO/European models.[3] Making any *a priori* assumption about the role of public authorities in constructing the authority of standards is of little use overall, as it depends on evolutionary variations regarding the political economy of state– market relations as much as on preferences regarding the issues concerned (Dudouet et al., 2006: 389). This is noteworthy with services, which can be highly technical, but at the same time embody contentious political interests and societal values. For instance, all sorts of standards related to information and communication technologies are used in services related to the development of smart global cities for improving transportation, energy efficiency, sustainable planning, and so on, but none of them would be of much use if left in a regulatory vacuum. More generally, the multiplicity of standards surrounding our everyday life has an influence on our health and safety, regardless of their place in regulation. As we saw in the previous chapter, the inclusiveness of standard-setting processes remains an issue whose significance lies beyond mere regulatory public policies. In a nutshell, standards are regulation.

This prompts us to have a closer look at the institutional settings for service standards provided by the activities of formal SDOs within the ISO environment, the European Union, the United States, as well as by the prospects arising from the new generation of trade agreements.

[3] For further detail, see Graz and Hauert (2014).

The ISO Setting

The ISO is a major arena for assessing current developments of service standardisation. As the world's largest developer and publisher of international standards with a membership of 160 or so mixed private and public national standardisation bodies, the ISO represents the wide range of public and private actors involved in services standardisation. The move into standardisation of services began in 1995 with a Consumer Policy Committee (COPOLCO) workshop in Beijing. Lawrence D. Eicher, then ISO Secretary General, emphasised that manufacturing industry was already changing with the move into generic management system standards and, from there on, 'the emphasis could change even more to take into account the needs of the burgeoning service industries' (International Organization for Standardization, 1995). Six workshops were held in the following years with various foci, such as tourism, exhibition management, banking and insurance, and engineering consultancy, as well as multi-sectoral methodological issues for developing service standards. In 2001, a new working group was established to draft a guide on the use and development of service standards from a consumer's perspective (ISO/IEC Guide 76:2008, Development of Service Standards – Recommendations for Addressing Consumer Issues). Since then, service standards not only appear each year as a key priority area of the work programme of the COPOLCO; this also led the ISO to develop its own Strategy for Service Standardization (International Organization for Standardization, 2016a) to increase ISO's visibility in this domain, help members develop service standards, and better understand market interests.

The number of ISO standards in relations to services is 700. This is still few (approximately 3 per cent) compared to more than 22,000 international standards and standards-type documents in the whole ISO catalogue (International Organization for Standardization, 2017: 5). Moreover, such figures should be taken with caution as they not only refer to specific requirements to be fulfilled by a service but also all sorts of standards that can support service provision.[4] Thus, standards labelled as belonging to services include domains far removed from what is usually understood as services, such as transport infrastructure, lab techniques, and construction engines. The broad inclusiveness of the international classification for standards shows the uncertainties in defining and classifying service standards, which can never be taken for granted. Yet some developments have taken place in domains

[4] For further detail on such a distinction between services standards and standards for services, see CEN (2017: 12ff).

epitomising core intangible and relational features of services. This is particularly the case for professionals providing personal financial planning such as in pensions per capitalisation (ISO 22222:2005), in the vocabulary and service requirements for market, opinion, and social research (ISO 20252:2012), and in safety requirements for scuba diving (ISO 24801–1:2014), as well as minimum quality requirements for services provided by tourist information offices (ISO 14785:2014). Those distinct sectoral standards remain marginal in terms of the global service economy. Obviously, large parts of this economy, such as finance and insurance, use instruments developed within their own sector, even if their ability to legitimately claim great authority in self-regulation has been seriously challenged in the context of the global economic crisis. Cross-border service providers also rely on more generic standards, which may indifferently be applied in the production and exchange of goods and services. Among the most widely used are the quality, environmental, and information security management system standards ISO 9000, 14000, and 27000 series, as well as the guidance on conformity assessment provided by the ISO 17000 series or the ISO 31000 guidelines and principles of risk management (Guler et al., 2002; Prakash and Potoski, 2006; Lalonde and Boiral, 2012).

Within the ISO setting, the development of service standards raises challenges pertaining to their content and the distinctiveness of services as compared to generic management standards. The relational and immaterial features of services prompt the development of standards that encroach simultaneously upon the intended quality of a service and the business operating procedures to deliver such a service. In the ISO, the latter is understood as a management system standard (MSS) and is kept separate from the former with dedicated procedures.[5] According to this so-called exclusion principle, any light quality management standards is ruled out from the back door. Yet, according to a number of participants in ISO technical committees, this may sometimes hinder the development of services standards or diminish their attractiveness to end users (International Organization for Standardization, 2017: 12). MSS represents a highly sensitive field of standardisation that requires a so-called justification study (formerly known as ISO/IEC Guide 72) and the adoption of a common document structure and terminology.[6] For

[5] See ISO/TMB Resolution 18/2012, available at: http://isotc.iso.org/livelink/livelink/fetch/-15620806/15620808/15623592/15788626/TMB_Communiqué_Issue_Nr._40_%28March_2012%29.pdf?nodeid=15787295&vernum=-2, accessed 18 October 2013.

[6] The recent revisions of the *ISO/IEC Directives, Part 1, Consolidated ISO Supplement — Procedures specific to ISO* (eighth edition, 2017) were precisely intended to rule and harmonise the development of management system standards with the introduction in

instance, the distinctiveness of services and the desire of small and medium enterprises to refer to one single standard as a reference has led the ISO technical committee on tourism to send several requests to the ISO governing body (the Technical Management Board-TMB) asking them to reconsider these rules. In 2012, these requests were unequivocally refused by the TMB, leading to substantial modification and adding to the existing delays in the development of service standards in the tourism sector.[7] Such requirements have impeded the development of service standards in many domains. Overcoming this difficulty will only be possible by setting standards according to a very narrow understanding of the procedural and generic aspects of services. This will make it difficult to include more substantial issues related to societal values and cultural contexts affecting the co-production of services.

Almost fifteen years after the 2005 ISO workshop 'Global Trade in Services – New Challenges for International Standardization' and twenty-five years after the launch of the institutional process, progress within the ISO has been meagre. Whereas some developments, such as those in risk management (ISO 31000) or energy management systems (ISO 51000), may come to have a major impact on the service sector, so far, maturity in service standardisation remains weak within the ISO environment.[8]

The European Approach

More developments take place in Europe with the European Union in the forefront of both service integration and international standardisation.[9] In 1985, Council Resolution 85/C 136/01 on a 'New Approach' to technical harmonisation and standardisation instigated a completely new

the annex SL of a 'High level structure, identical core text and common terms and core definitions for use in Management Systems Standards'.

[7] See ISO/TMB Resolution 17/2012, 'Management Systems Standards in tourism and related services', available at www.iso.org/iso/copolco_priority-programme_annual-report_2012.pdf, accessed 18 October 2013.

[8] For instance, since the creation of the ISO committee on tourism service standards in 2005, almost half of the international standards published so far come from only one out of its ten working groups (in the recreational diving sector; in contrast, accessible and sustainable tourism, or health tourism services have hardly progressed); see ISO TC/228 webpage for further detail: https://www.iso.org/committee/375396/x/catalogue/p/1/u/0/w/0/d/0 (accessed 3 August 2018).

[9] There are other regional standardisation bodies, most notably in the Americas (Pan American Standards Commission, COPANT and Asociación Mercosur de Normalización, AMN) and in Asia-Pacific (Pacific Area Standards Congress, PASC) and in Africa (African Regional Organization for Standardization, ARSO). As compared to the European system, however, their influence is still weak.

regulatory technique and strategy. The resolution was a response to the growing role of the European Court of Justice in resolving conflicts on regulatory policies in the internal European market, especially since the 1979 Cassis de Dijon case securing the principle of mutual recognition in the absence of harmonised legislation or technical standards. It was also an early move towards the completion of the Single Market by devising procedures to avoid turning technical specifications into structural impediments to trade. Although member states were suspicious about seeing regulation in this domain transferred to the European authorities, they did perceive the threat of a race to the bottom in public purpose standards as market integration progressed. The New Approach provided a framework for the harmonisation of EU public law only on the general and essential requirements of goods traded on the European market, in particular in the fields of health, environment, safety, and consumer protection. Depending on the sectors affected, technical specifications, performance criteria, and quality requirements are either based on mutual recognition of national standards or delegated to European standard-setting bodies upon formal request from the European Commission. In most sectors, the procedure for monitoring standards is a matter of business self-regulation, since products put on the market are granted a presumption of conformity, solely based on the declaration of the manufacturer (CE marking). Thus, the European New Approach has done more than strengthen the ability of companies to rely on voluntary standards rather than mandatory regulation in the Single Market. By avoiding costly third-party testing and certification, and providing the procedural means for a simultaneous adoption of European standards as international ones (through the so-called Dresden and Vienna Agreements), the EU has also included third countries in its standardisation system. The (largely unintended) outcome has been a powerful strategic positioning of European standards in the global market (for more detail, see Vogel, 1995; Egan, 2001; Borraz, 2007).

The European Commission was well aware that the emergence of an increasingly dense and extensive European standardisation complex with global reach could also support the 2000 Lisbon Agenda. Services were a core feature of the plan 'to become the most competitive and dynamic knowledge-based economy in the world'. New emphasis on service standards occurred after the 2005 mid-term review of the Lisbon Agenda and adoption of Directive 2006/123/EC on services in the Internal Market, the so-called Bolkestein Directive, eventually agreed to on second reading in December 2006 and fully implemented since the end of 2009. A horizontal approach to regulatory harmonisation supposedly valid for any kind of service provision at the European level lies at the centre of the directive. The controversial 'country of origin' principle has been substituted for the formula 'freedom to provide services' in order to ensure conformity with regulations of the place of delivery. The Directive emphasises that the promotion of quality is a crucial issue for the unification of the internal market for services. To this end, it explicitly encourages professional independent associations and standard-development and certification bodies (like the CEN, CENELEC, or ETSI) to develop voluntary quality marks and labels (preamble 102 and article 26).

Against this background, the European Commission undertook a series of action to support service standardisation. It addressed in 2003 a first Programming mandate (M 340) to CEN, CENELEC, and ETSI in the field of services to identify priority sectors of intra-community trade in services. Issues could include horizontal cross-sectoral generic standards and vertical sector-specific standards, as well as service providers or end-users. A second programming mandate (M 371) was addressed to CEN in the field of services in 2005 following the transfer of responsibility for business-related services to DG Internal Market and Services. Half a dozen European standardisation bodies developed eleven projects accordingly. It is worth looking at them in some detail, as the result of this whole exercise shaped the new Regulation on European Standardisation (1025/2012), adopted in October 2012.

The CEN Horizontal European Service Standardization Strategy (CHESSS) was the largest project responding to EU Mandate M/371. It included a consortium of national standards bodies led by the British Standards Institution (BSI), with those from Spain (AENOR), Germany (DIN), Denmark (DS), Estonia (EVS), and the Netherlands (NEN), as well as CapGemini, one of the world leaders in IT services consulting and management. Its final report, published in 2009, examined the feasibility of a generic approach to European service standardisation across multiple service sectors, as opposed to following a sector-specific approach (CHESSS Consortium, 2009). The CHESSS project has raised crucial issues on the distinctiveness of service standards, echoing the aforementioned discussion regarding ISO's 'exclusion principle' between quality management standards (i.e. ISO 9000 series) and what can be purposely standardised for the service sector. Indeed, the importance of quality in services inevitably led to questioning their specificity with regard to quality management standards. One module of the project claims that service standards are not about the 'how' but about the 'what', i.e. a service standard is not about how to achieve a goal, as with management standards, but specifies the goal to be achieved and the means for assessing its achievement (CHESSS Consortium, 2009, modules 4 and 5). In this regard, the proposal to develop a customer satisfaction index is undoubtedly as crucial in framing conformity assessment procedures in services as weights and measures underpinning similar procedures for products. The distinctiveness of service standards is that they extend beyond procedural issues to cover such issues as common writing models and the terminology employed across the entire service sector. The CHESSS project clearly aimed at ensuring that service standards establish their distinctiveness in the realm of standardisation, as management or performance standards did previously.

Besides the distinctiveness of service standards as such, unsurprisingly, the CHESSS report pointed out the difficulty of involving stakeholders in the development of generic standards when most of them lack the necessary awareness and resources. The difficulty of stakeholders' involvement in service standardisation has not just been rehashed time and again in subsequent reports.[10] It also casts doubts on the institutional structures for setting service standards. For some, the current system is as appropriate for services as for products. In contrast, B2B services are seen as a good case for a new system based on a dual representation with stakeholders besides conventional national bodies, such as European organisations representing industry, SMEs, and consumers. According to the CHESSS report, 'This double representation system ensures a balanced representation of sectors on the one hand and of national interest on the other hand' (CHESSS Consortium, 2009: 223 (module 7)). Such recommendations have struggled to gain a significant place in subsequent European initiatives in the wake of the 'Standardisation package' adopted by the European Commission in 2016. However, the interest in a single horizontal generic standard with a certification scheme is clearly an attempt to promote service standards on a par with the worldwide achievement of the ISO 9000 series. Thus, the important role of the European Commission in supporting standardisation for the service sector may not only reinforce endogenous recognition of service standards. It could also pave the way for the deterritorialisation of regulatory practices through greater reliance on market mechanisms for the diffusion of such standards.

In contrast, the ten other projects responding to EU Mandate M/371 addressed the specificity of distinct service markets. Afnor, the French national standardisation body, a pioneer in setting national standards in well-defined service sectors, initiated those projects in consultation with some European partners, in particular from the Netherlands and

[10] See in particular European Commission (2016e).

Denmark. The recommendations identify a number of service activities likely to be standardised at various levels, whether European Standards per se, or at a lower level, guidance materials and so-called workshop agreements.[11] The advantage of a vertical and sectoral approach is largely seen in the quality of the deliberation process likely to better address the distinctiveness of services in sectors of highly relational and immaterial activities. According to Pascal Gautier, head of the Management and Services Unit at Afnor, generic standards in services would soon become burdensome and unrealistic as 'they require phenomenal efforts which would eventually generate opposition'; in his view, 'it is much better to favour a niche approach in service standards so as to keep a sector-specific proximity, i.e. to choose a so-called Swiss army knife effect where each blade has its distinct use'.[12] However, the ambiguous mixture of private and public actors involved in standardisation processes favoured by this approach remains important. Similarly, the issues concerned do not clearly distinguish between societal or more strictly technical objects of reference. A proper differentiation of actors among stakeholders and issues spanning physical measure to societal values, as well as clear-cut incentives to mitigate representation biases, would be necessary to ensure a fair, substantial, and thorough representation in standardisation processes.

In the wake of these early moves, the Commission initiated a reform of the European standardisation system.[13] Faced with the faster development of service standards at the national than at the European level, the potential creation of barriers to intra-EU trade in services, and services increasingly embedded in the delivery of goods, one of the key objectives was to establish a better inclusion of service standards in the regulatory framework. Despite opposing views of what can be standardised in services, the

[11] CEN/CEN Management Centre, *Summary, Background and Proposals related to European Commission Programming Mandate M/371 in the Field of Services* (n.d. April 2009). According to the report, standardisation work should be initiated in the following areas: accessibility of transport and tourist services, project management services in the field of engineering consultancy, services for residential homes and older persons, reception services, IT- and non-IT service outsourcing, and smart house services.

[12] Author's interview with Pascal Gautier, Head of the Management and Services Unit, Afnor, Paris, 18 April 2007.

[13] See for instance COM 2011(311) Final: 'Progress in the development of European standards for services has, however, been slow and recent years have seen the rapid growth in service standards at the national rather than the European level, (453 new national standards in 2005–2009, as opposed to only 24 European).' The reform has incorporated Directive 98/34/CE of the European Parliament and of the Council regarding the 'procedure for the provision of information in the field of technical standards and regulations and of rules on Information Society services' (22 June 1998) and Decision 1673/2006/CE of the European Parliament and of the Council on the financing of European standardization (24 October 2006).

consultation organised in 2010–2011 led to strong support for including service standards and keeping the principle of national delegation in this domain.[14] As such, the entry into force in 2013 of the new regulation on European standardisation (1025/2012) extended the New Approach to services and compelled European national standardisation bodies to provide notification of services standardisation activities. Moreover, the new environment reinforces the support granted to European civil society stakeholders and SMEs. Nonetheless, the new regulatory framework has not necessarily diminished the divide opposing supporters of vertical sector-specific standards, such as Afnor, and horizontal cross-sectoral generic standards, such as those promoted by the British Standards Institution (BSI). This probably explains the mid-range strategy pursued by the European Commission in addressing Mandate M/517 in January 2013 to the CEN, CENELEC, and ETSI for the development of 'horizontal service standards'; while fostering the standardisation of the generic attributes of services, the mandate emphasises the development of '"narrower" horizontal service standards for particular aspects/parts of a full service provision' as opposed to a single, all-inclusive horizontal service standard. As a result, the framework devised by the European Commission for the development of European service standards explicitly includes the option of a 'hybrid combination: a horizontal service standard with sectoral add-ons, or a pool of parallel sector-specific standards' (European Commission, 2016c: 9). Regarding horizontal service standards as such, out of six topics identified by the European standardisation organisations as suitable for future developments at the European level, the European Commission retained the following three, explicitly listed in the new technical committee CEN/TC 447 'Services - Procurement, contracts and performance assessment' created for that purpose in 2016.

The United States: A Special Case

With European standardisation processes usually seen as driven by a coherent and centralised institutional framework in opposition to the fragmented and decentralised American system, the overall design of international standardisation remains unsurprisingly disputed. In Mattli's words,

[14] See: http://ec.europa.eu/enterprise/policies/european-standards/standardisation-policy/policy-review/public-consultation-2010/index_en.htm, (accessed 14 April 2015). Documents adopted in June 2011 by the European Commission are the following: Communication on a strategic vision for European standards – COM (2011) 311; Proposal for a Regulation on European Standardization – COM(2011)315. These communications extensively draw from the following report: EXPRESS (2010).

the disagreement between Europeans and Americans is about whether an international standard is simply one that benefits from *de facto* or *de jure* international acceptance and use by an industry, or whether it must come from an organisation that is truly international in the sense that it has an international representation of national members and an international voting structure based on those national members. A resolution of this disagreement is not in sight; it will require, among other things, a clearer understanding of the relationship between national, regional and international standardization organizations. (Mattli, 2001: 330).

As seen previously, despite a number of noticeable differences between the American and European systems, several features do contribute to bridging the conventional gap of such a transatlantic divide.

First, the American National Standards Institute (ANSI), a not-for-profit private organisation, plays a significant role as the national standardisation body in centralising standardisation processes. Its mandate explicitly places ANSI in charge of the coordination and representation of US interests at the ISO and IEC. ANSI also plays a crucial role in enhancing the coherence of standard-setting processes both within the United States and amongst US participants in international arenas. Without developing standards, it coordinates and accredits US-based SDOs, which in turn must comply with the ANSI essential requirements for standards development processes. In fact, the Vice President of International Policy of ANSI takes issue with the depiction of the American standardisation processes as fragmented and decentralised: according to him, they take place in an 'organised distributed system'.[15] This particular account of the American setting presents ANSI's coordination role in a positive light but also reflects the delineated environment in which US standardisation takes place.

While ANSI is responsible for the coordination of over 200 accredited SDOs, the National Institute of Standards and Technology (NIST) is the federal agency that fulfils a similar role at the level of governmental agencies. Over recent decades, the 1996 National Technology Transfer and Advancement Act and successive revisions of the Circular A-119 of the Office of Management and Budget have played a significant role in enhancing NIST's profile. Those pieces of legislation and regulation entrust NIST with promoting the use of voluntary standards in lieu of government-unique standards within federal agencies. Whenever government-unique standards are used, they must be fully reported and justified. Moreover, staff across federal agencies take part in the development of voluntary standards in over 500 SDOs, with personnel

[15] Interview with Gary Kushnier, Vice-President for International Policy, ANSI, Washington DC, 7 August 2009. Note: all interviews in Footnotes 15–23 were carried out by my research assistant Christophe Hauert.

from NIST alone in 114 SDOs in 2016 (National Institute of Standards and Technology, 2017; United States Government Accountability Office, 2018). The direct involvement of governmental agencies is only part of the relationship between public authorities and standardisation. More than 8,600 standards are referred to in US law, and over 10,500 in public procurement procedures. It is also worth noting that the ANSI Steering Committee not only includes representatives of industry and civil society but also a number of government agencies.

More generally, the primacy of international standards is explicitly recognised in the American system, even though the understanding of what 'international' means still remains controversial in the *United States Standards Strategy* (USSS). This remains, despite the substantial USSS revision passed in 2005, specifically to address such needs and more recent suggestions made by the Government Accountability Office under the aegis of the Trump administration that NIST should better 'respond to circumstances when U.S. representation in international standards activities may be inadequate' (United States Government Accountability Office, 2018: 53). Last but not least, as in European reforms, the American system has recognised that participation from the weakest stakeholders is in such short supply that it undermines the legitimacy of technical specifications supposedly driven towards the public interest; this is why recent policies on both sides of the Atlantic have taken initiatives supposedly geared towards supporting the participation of civil society organisations. The US standards strategy points out that 'government should recognize its responsibility to the broader public interest by providing financial and legislative support, and by globally promoting the principles of our standardization system' (American National Standards Institute, 2016: 12). In brief, American standardisation processes rely on a broader mix of public and private actors than usually acknowledged.

While the American picture is not dissimilar to the European and ISO ones, current developments in the distinct domain of service standards remain sharply dissimilar across the Atlantic. Services are for the most part low on the agenda of American SDOs. Even the largest standard-setters pay scant attention to how services may challenge the future of standardisation. The American Society of Mechanical Engineers (ASME) includes clean energy and robotics among the five core technologies targeted by its latest strategic plan, to which a number of services could potentially be associated.[16] Yet none of them specifically

[16] American Society of Mechanical Engineers, *ASME Strategy, approved by ASME Board of Governors May 2018*, available at https://www.asme.org/wwwasmeorg/media/ResourceFiles/AboutASME/ASME_Strategy-180614.pdf (accessed 3 August 2018).

focuses on services. As Bernard Hrubala, Vice-President of ASME and Division Manager at TÜV Rheinland, put it when questioned about a distinct service strategy, 'our ultimate goal at the end of the day is, don't matter what the standard is in every country, we want their standards to be consistent with the ASME standards'.[17] ASTM International (originally known as the American Society for Testing and Materials) shares this claim to play a leadership role at the global level with an active policy of memoranda of understanding signed with more than one hundred national standards bodies, mostly in developing and emerging countries (Saudi Arabia, Columbia, and Turkey being the three countries most referencing those standards). Yet it ignores the issue of service standards and prefers to give prime importance to sustainability. It is from this standpoint that ASTM International has revised most existing standards and charts new activities such as carbon footprint and alternative fuels. Several years ago, Katharine E. Morgan, who is now President of ASTM international, went to great lengths to explain this shift: 'We are seeing green, from roofing to isolation to degradable plastics, we are seeing that across a lot of our committees'.[18] For its part, NIST sees its role in service standardisation as closely related to strategic issues set by the US administration in domains closely related to recent advances in computing, communications, defence technologies, and healthcare (National Institute of Standards and Technology, n.d.). Finally, in 2013 ANSI launched a Services Sector Initiative to help meet the demands of standardisation within the US services sector and identify priority sectors. While the recommendations made in the wake of this initiative repeat claims towards greater awareness, visibility, outreach, and engagement, it also suggests a need to 'identify common elements that cut across all service sectors not just one or two specific sectors'.[19] Even if the initiative has so far merely led to a few conferences and an enhanced dedicated website, this clearly contrasts with earlier views, according to which abiding by its coordination mandate would be at odds with setting any priority at all as long as its members have not done so – which *de facto* excluded service standards.[20]

17 Interview with Bernard E. Hrubala, Sr., Vice President, ASME, and Division Manager of Industrial Services, TÜV Rheinland, New York, 18 August 2009.

18 Interview with Katharine E. Morgan, Vice President, Technical Committee Operations, ASTM International, West Conshohocken, 19 August 2009.

19 Services Sector Initiative Summary, PPT presentation for the ANSI–ESO Meeting, February 21–22, 2017. Available at: https://share.ansi.org (accessed August 3, 2018).

20 Interview with Gary Kushnier, Vice-President for International Policy, ANSI, Washington DC, 7 August 2009.

Overall, standardisation in services does not lie at the heart of the American landscape. Interestingly, the few service standards dealt with among American SDOs are confined to domestic issues. For instance, the development of the ASTM Environmental Site Assessment Standard (E1527) merely responded to a requirement set by the US Environmental Protection Agency.[21] Officials in charge of standardisation strategy in the major bodies of the American institutional setting invariably explain the weak concern over service standards by a lack of demand. In ASME words: 'Our scope is essentially mechanical engineering. Those services type things don't really fall within our area.'[22] Moreover, service standards raise the issue of certification. American SDOs remain highly critical of standards likely to be used for certification purposes. Taking the example of the ISO 9000, ANSI emphasises the lack of added value brought by certification: 'It didn't add value if you are a large company and you already have an excellent quality management system. What does it bring to spend a few more millions of dollars or euros to get certified to something you do better already?'[23] Whether it be an aversion towards certifiable standards or merely qualified isolationism, such a view may face renewed challenges in the importance recently taken by regulatory convergence in negotiations to establish far-reaching free trade agreements across the Atlantic and the Pacific.

Towards New Transatlantic and Transpacific Promises?

During the confirmation hearing before the European Parliament for her appointment as Trade Commissioner, Cecilia Malmström repeatedly emphasised the 'strategic dimension to the regulatory work'. Referring to the contentious negotiations under way between the European Union and the United States for the establishment of a Transatlantic Trade and Investment Partnership (TTIP), she claimed that '[i]f the world's two biggest powers when it comes to trade manage to agree standards, these would be the basis for international cooperation to create global standards'(European Parliament, 2014: 8). Similarly, in a brochure listing what the EU Trade Commissioner saw as ten myths about TTIP, the strategic dimension of setting high standards in global trade was the first point in

[21] Interview with Katharine E. Morgan, Vice President, Technical Committee Operations, ASTM International, West Conshohocken, 19 August 2009.

[22] Interview with William Berger, Managing Director, Asme, and Bernard E. Hrubala, Sr. Vice President, ASME, and Division Manager of the Industrial Services Unit, TÜV Rheinland, New York, 18 August 2009.

[23] Interview with Gary Kushnier, Vice-President for International Policy, ANSI, Washington DC, 7 August 2009.

countering the idea that TTIP would weaken strict EU standards to protect people and the planet (European Commission, 2016b). Whether it be the aborted TTIP, the Comprehensive Economic and Trade Agreement (CETA) provisionally entered into force in September 2017 between Canada and the European Union, the EU-Japan Economic Partnership Agreement entered into force in February 2019, or the Comprehensive and Progressive Agreement for Trans-Pacific Partnership (CPTPP) signed in March 2018 between eleven countries of the Pacific rim, the authority conferred on technical standards has swiftly become one the most prominent issues along with investor-state dispute settlements mechanisms in negotiations of a new generation of preferential trade agreements.[24] European fears about the abolition of food safety standards as protection against importation of 'frankenfood' such as chlorinated chicken or hormone beef from the United States may be exaggerated. Still, the conflict-ridden negotiations for such mega-trade deals do include provisions for new harmonised standards (including in services), greater mutual recognition of existing standards, and plans to set up joint councils in charge of designing future convergence around the type and scope of technical standards to be recognised in government regulations.

The momentum towards this new generation of free trade agreements is unmistakably facing setbacks in the wake of the strong opposition coming from both the right and left of the political spectrum. Indeed, as soon as Donald Trump acceded to the United States presidency in January 2017, he pulled out of both TTIP and the Trans-Pacific Partnership and demanded a renegotiation of the North American Free Trade Agreement (NAFTA) with Canada and Mexico. A number of demonstrations and election campaigns in Europe have put the case against such agreements at the core of their demands, notably during the 2017 French Presidential elections. Also worthy of note is the tortuous ratification process by as many as thirty-eight national and regional parliaments for CETA. It demonstrated the increasing politicisation of a number of trade-related

[24] In the wake of Brexit, the United Kingdom's decision in the 23 June 2016 referendum to leave the European Union (EU), negotiations to establish new trade agreements will certainly deal with similar issues as well. Moreover, mention should be made of another large preferential trade agreement under negotiations at the time of writing: the Regional Comprehensive Economic Partnership (RCEP) between the ten member states of the Association of Southeast Asian Nations (ASEAN) and the six states with which ASEAN has existing free trade agreements (Australia, China, India, Japan, South Korea, and New Zealand). If signed, the agreement will concern a combined population of 3.4 billion and trade volume accounting for nearly 30 per cent of the world's total trade; while such figures are certainly massive, the agreements only make limited demands on regulatory and standards convergence besides intellectual property rights (Ravenhill, 2016).

issues, including standards and regulatory convergence, as if such a thing had ever been in doubt. When Paul Magnette, the minister-president of Wallonia, stood alone in a tough constructive critique of CETA to force his European and Canadian counterparts to take his concerns seriously as a condition for the required ratification by the regional Parliament of Wallonia, he was certainly right in claiming that 'This is not only about a treaty with Canada; this is about all future bilateral [trade] agreements. The question actually is: which globalisation do we want?'[25]

With CPTPP signed without the United States and its ratification still underway at the time of writing, TTIP in limbo, and EU partership agreements such as those with Canada and Japan only recently entered into force, it is not worth making a detailed study of the outcome of such negotiations with all their sector-specific variations. It is worthwhile, however, to seek a better understanding of how those new preferential trade agreements are likely to confer authority on standards as de facto or de jure regulatory instruments governing further internationalisation of services. According to the few independent and in-depth studies that exist so far, an ambitious harmonisation of standards would no longer be on the agenda (Cai, 2016; De Ville and Siles-Brugge, 2016: 38–61; Pitschas, 2016; Magnette, 2017). Even if negotiations on mega-trade agreements resume in the future, the prospect of setting global standards thus remains unlikely or at least questionable. Any likely outcome would turn on a mutual recognition of existing standards rather than their harmonisation. Four issues are at stake in this regard.

The first concerns the regulatory chill effect that far-reaching preferential trade agreements may have on governments. This chilling effect characterises a situation in which governments become reluctant to adopt new regulations or to strengthen existing standards for fear of scaring off market actors, particularly foreign investors. The risks of a regulatory chill effect at the expense of welfare policies, consumer protection, public health, or environment policies have chiefly been discussed in connection with negotiations on new investor-state dispute mechanisms, rather than the regulatory cooperation chapters under which harmonisation and mutual recognition of standards are negotiated. They are therefore beyond the scope of this study. They may, however, just as well result from a harmonisation or a comprehensive mutual recognition of standards (De Ville and Siles-Brugge, 2016: 79ff). It is plausible to consider that the sovereign right to uphold a level of

25 Quoted in 'La Wallonie dit toujours non au CETA', radio television belge francophone (online), 21 October 2016 (my translation) https://www.rtbf.be/info/belgique/detail_ceta-paul-magnette-s-explique-en-commission-du-parlement-wallon?id=9436240.

protection would be seen as useless when lower levels would be just as acceptable by means of mutual recognition. This is particularly the case where regulations are not completely equivalent in terms of outcome. An outcome in levels of protection less ambitious than the status quo could also result from a joint adoption of less stringent international standards, as for instance with the Codex Alimentarius, as compared to a number of provisions included in European sanitary and phytosanitary standards. Finally, the chilling effect can arise from regulatory cooperation procedures devised for setting future standards. For instance, CETA includes provisions to 'discuss regulatory reform and its effects on the Parties' relationship', with guarantees that 'consultation and exchange should begin as early as possible' in regulatory development processes (European Commission, 2016a, art. 21.4.a(i), (art. 21.4.b)). The very fact of having the duty to consider the effects of regulatory reform in the Parties can cool down eagerness towards new or more stringent standards. Moreover, in the absence of dedicated mechanisms to support the involvement of civil society organisations irrespective of their resources, such provisions may unduly benefit business organisations with privileged access to this type of consultation mechanisms and often more reticent towards new or more stringent standards. This brings me back to the question seen again and again in the course of this book: who sets the standards?

The second point raised by the new generation of preferential trade agreements is indeed the transfer of authority in standard-setting procedures likely to flow from their mechanisms of regulatory cooperation. Unlike previous treaties, such agreements are designed as 'living agreements', where parties can engage in new areas of regulatory cooperation without the need to re-open the initial international agreement or to modify each other's institutional framework (Alemanno, 2015: 631–632). The implementation of future regulatory convergence may thus take place outside existing regulatory agencies related to sovereign states or the EU. Be it the Committee on Regulatory Coherence imagined for CPTPP or the Regulatory Cooperation Forum established by the CETA, this raises significant concerns when the time comes to define in more detail the membership, scope, and functioning of the bodies established for the purpose of such on-going regulatory cooperation. Two issues stand out in particular. First, regarding membership, considering the influence that such bodies may have on future regulation, a fair and balanced representation in defence of the public interest is particularly important. At first sight, CETA, as a forerunner of potential future agreements, appears unambiguous in this regard. It holds that the Regulatory Cooperation Forum 'shall comprise relevant officials of each Party', i.e. from regulatory authorities (art. 21.6.3). Yet it directly adds that 'other interested parties to participate in the meetings' may as well be invited by mutual consent. While this is clearly consistent with consultations with private parties such as representatives from academia, think-tanks, NGOs, businesses, and consumer and other organisations (article 28.8), here again it leaves the door wide open to distorted lobbying practices as long as it does not include provisions and provide public support to make up for the over-representation of well-resourced business organisations (with less detailed language, CPTPP articles 25.6 and 25.8 raise similar concerns). The second issue deals with the mandatory or voluntary nature of such cooperation. Taking again the case of CETA as the only agreement ratified so far in this domain, it holds that regulatory cooperation activities are undertaken 'on a voluntary basis', but requires parties to provide explicit explanations in case it refuses to initiate regulatory cooperation or withdraws from cooperation (article 21.2.6). Such a burden of justification against the voluntary principle was seen by the Wallonian Parliament as crucial enough in the course of its contentious ratification process that it insisted on including the following plain language in the Joint Interpretative Instrument added to the signature of the agreement: 'regulatory authorities can cooperate on a voluntary basis but do not have an obligation to do so, or to apply the outcome of their cooperation' (Council of the European Union, 2016, §3). Without such plain language, there would indeed be more of a place for imposing ever increasing areas for convergence to become legally binding.

A third and much-discussed issue is the potential outcome of a greater mutual recognition of existing standards. As Vogel (1995) forcefully argued more than two decades ago, increased economic integration is not necessarily incompatible with stronger regulation and standards in domains such as labour, environment, and consumer protection. Yet, as seen earlier, mutual recognition is more likely to lead to a race to the bottom than to the top, as regulations are rarely completely equivalent in terms of outcome. This might also be the case with services such as education and training, engineering, architecture, electronic communications, transport, legal services, and so on. In contrast to goods, regulations and standards for services are often more decentralised and set by sub-state or non-state bodies such as professional association or private entities – and therefore more difficult to compare. Moreover, service regulations and standards rarely concern the service itself. In order to respond to no end of quality and security issues prompting market uncertainties, they are more likely to define conditions that service providers are expected to fulfil (professional qualifications, etc.) or the circumstances of the services' delivery (opening hours, location, safeguarding of public services, etc.). As a result, they tend to be more diversified and complex than those pertaining to goods, making their equivalence even more difficult to assess in mutual recognition procedures. Think of someone who completed a professional degree in country A and then moves to country B to take up a job, where she finds that authorities of country B don't recognise her diploma and ask her to pursue two additional years of study before she could apply for the job which she could have directly taken up in her own country.

At the same time, as service regulation and standards concern the process and, as such, are not necessarily reflected in the actual 'content' of a service, they may well be seen to have less effect in the importing country. This explains why importing hormone beef raises more concerns than software programmed in hazardous conditions, even though the latter could breach key information security and protection of privacy requirements. As Hatzopoulas points out, service regulation and standards will 'meet much lower resistance from consumers in the host State – and therefore local [...] rules will be under greater pressure from regulatory competition. If the host State is to safeguard its own standards ... it needs specifically to legislate acts of an essentially protective nature. Such rules are unlikely to yield to the effect of mutual recognition' (Hatzopoulos, 2012: 63). In brief, even if we accept the oft-repeated discourse that no provision in mega-preferential trade agreements under negotiation would whatsoever lower existing levels of protection, extending the principle of mutual recognition to service regulations and standards such as professional qualifications (e.g. CETA chapter 11), licensing requirement, and approvals procedures (e.g. CETA chapter 12) is at best intricate and at worst might well be detrimental. At the time of writing, the present state of pending or discontinued negotiations and the types of provisions included in the negotiations achieved with CETA do not allow for clear conclusions.

A final point concerns the implications of mutual recognition of existing regulations and standards for third countries not part of the preferential trade agreement. Irrespective of provisions agreed or under negotiation, preferential trade agreements must comply with the rules established by the WTO. Legal scholars concur that, while the WTO framework is not entirely clear, or coherent, it provides a rather open understanding of how recognition agreements should avoid discriminatory implications for third-parties (Trachtman, 2003; Nicolaidis and Shaffer, 2005). Regarding services, GATS article VII.2 sets out that States 'shall afford adequate opportunity for other interested [States] to negotiate their accession to such an agreement or arrangement or to negotiate comparable ones with it'. In its attempt to respond to the fundamental principle of most-favoured-nation (MFN), on which the whole architecture of non-discrimination rests, GATS article VII.2 clearly does not prompt any automatic extension of mutual recognition agreements to third parties. Yet, by means of procedures of notification and ensuing participation to negotiations under way, Mathis (2012: 72) points out that 'MFN plays at least a 'conditional' role to assess the potential participation of third parties.' Here again, such principles may be more difficult to realise with services than with goods. As recognition in the domain of services mainly concerns professional qualifications, licensing requirements, and approval procedures, they are more likely to be granted to individuals and firms on a one-by-one basis – in contrast to products, whose conformity assessment is more likely to be valid to all the same products put on the market. Accordingly, it does not require in-depth legal expertise of the provisions negotiated in the context of the new generation of preferential trade agreements to realise that parties who will not be part of the mutual recognition provisions designed for services in such agreements should not expect many spill-over effects for their own benefits, notwithstanding the relatively open WTO framework on mutual recognition of regulations and standards towards third-parties. Ultimately, as VanDuzer (2012) points out, the implementation of such intricate provisions eventually depends on local contexts as well, with actors such as domestic bodies, regulators, and sectoral experts being the real players engaged in the process. While this may add additional uncertainty to rules already identified as unclear and not always coherent, the discretionary power of local agents emphasises the ambiguous authority on which the recognition of standards and regulation rests against the background of the new generation of free trade agreements. Such ambiguity applies not just to third countries not part of those new mega-trade deals. As seen previously, it also supports the regulatory chill effect they may have on participating governments, their mechanisms of regulatory cooperation, and a mutual recognition of existing standards based on a loose understanding of equivalence.

Service Standards and Institutional Ambivalences

The following discussion focuses on how the aforementioned developments matter in assessing the authority of international standards in the service sector along the three core dimensions of my analytical framework, i.e. the institutional continuum of the actors involved, the material continuum of the issues concerned, and the spatial continuum along which such standardisation processes are likely to be recognised across jurisdictions.

Unsurprisingly, public and private actors very much overlap in the standardisation arenas on both sides of the Atlantic, as well as on the international plane of the ISO system and preferential trade agreements. There is also strong evidence of significant public support, in particular within European institutions. However, the limited results of initiatives taken over the years shed light on a common feature on both sides of the Atlantic as well as within the ISO and the context of the new generation of preferential trade agreements: the support and expertise of private actors is crucial in the development of standards. The low level of involvement in the field of service standards in the United States mirrors the difficulty of European and ISO projects in convincing stakeholders from the private sector. This suggests that behind labels of 'direct participation' in the United States and 'national delegation' for the European and ISO setting, actors setting standards are the same: large firms dominate technical committees, with government agencies attempting in some cases to take part in drafting standards, and not-for-profit associations from civil society remain largely under-represented.[26] The entry into force of EU Regulations 1025/2012 introduced new processes that improved the monitoring and participation of stakeholder organisations representing consumers', workers', SMEs', and environmental interests. Yet, the first evaluation undertaken under those new commitments points out, euphemistically, that such participation 'is still challenged', notably because of their weak position and different capacities in terms of stakeholders' representation at national level, as well as a lack of inclusiveness at the international level when standards are jointly drafted with ISO or IEC in the lead (European Commission, 2016d: 4, 11–12). In contrast to the direct political influence of the European setting, the American system relies on the indirect influence of the legal and regulatory environment supporting and legitimising the output of formal and informal SDOs. Thus, far from mere fragmentation, the US system hinges upon double coordination mechanisms, ensured by ANSI at the level of formal SDOs and by NIST with regard to governmental agencies. The distinction between national delegation and direct participation therefore appears to be more relevant for describing the space of standards recognition outside the United States than the type of actors involved within the United States. Finally, the difficulties experienced by the European attempts to foster the development of standards in the field

[26] For further detail on how global corporations able to set standards in their own interests, see: Graz (2018).

of services show how the enrolment of private actors can become an important political issue. The lack of distinct service standardisation processes in the United States here echoes the difficulty in encouraging stakeholder involvement in European projects to develop standards in conjunction with the unification of the market for services. Apparently, European officials have greater influence on the issues put on the agenda than do private actors likely to shore up the processes of setting new service standards. Defining the membership, the scope, and the functioning of the bodies established for on-going regulatory cooperation in the new generation of preferential trade agreements raises the same concerns.

Regarding the objects concerned, the potential scope of international standardisation in the domain of services differs greatly across the Atlantic and beyond. The antagonism between horizontal and vertical standards reflects the struggles at stake in defining what should be standardised in services: should it be the functional attributes of technical interfaces supporting the interaction between providers and customers on a horizontal basis for the widest range of services (information requirements, billing, complaint handling, etc.)? Or should technical specifications be more substantial on a narrower sectoral basis, defining how services can be co-produced and used on a reliable basis with shared expectations regarding their quality? Services' distinctiveness is clearly at stake here, with an assumption that the more intangible and relational the service is, the more difficult to measure, qualify, and standardise. Yet this does not mean impossible. In spite of all their flaws, European initiatives have helped build a coherent framework for the standardisation of services. The 'hybrid combination' imagined by the European Commission for the development of horizontal service standards with sectoral add-ons, or for a pool of parallel sector-specific standards, may eventually overcome the controversy between vertical and horizontal service standards. Even the more shallow horizontal approach may gauge the quality of services, Standards on performance measurement, service contracts, and service procurement expected from the Technical Committee (CEN/TC 447) established in 2016 could provide evidence of the positive impact of a standard on consumers. Moreover, it is worth noting that the case of energy and smart metering suggests that societal issues are likely to be greater in Europe than in the United States, where the focus is on narrower technical and market-driven aspects. While both sides demonstrate interest, American stakeholders narrow it down to technical issues associated with the physical characteristics of the resources delivered by such services. In contrast, European initiatives explicitly point out broader concerns of sustainable development, notably in relation to the implementation of the EU Directive on energy efficiency (2012/27/EU). Finally, this concerns how new standards and mutual recognition of existing ones are likely to lead to a race to the top. We saw that transatlantic and transpacific promises to set new global standards are greatly exaggerated. We should rather double-check the implications of extending the principle of mutual recognition to service regulations and standards such as professional qualifications, licensing requirements, and approval procedures. With the new generation of mega-trade deals still a moving target at the time of writing, the analysis can only be tentative. It shows, however, that such effects are at best intricate, but may well be detrimental.

This brings us to the third dimension defining the transnational hybrid authority of standards: the extent of the space in which technical specifications in the domain of services are likely to be defined, distributed, and recognised across sovereign States. International standards compete in terms of their different sources of legitimacy, as well as their various modes of cooperation. Market adoption is the main source of legitimacy for standards developed by American SDOs. This means that the recognition of standards beyond the sovereign space of the United States primarily relies on the exogenous process of market mechanisms – a good entry point into new markets as expressed by one interviewee. The translation of standards into official languages of various countries and the organisation of training workshops tailored to the distinct needs of well-chosen countries are an integral part of this strategy. This does not mean, however, that American SDOs overlook the legitimacy of their standards based on direct participation. The ASTM Memoranda of Understanding signed by more than one hundred national standards bodies strongly echo the principle of national delegation in use at the CEN and ISO, even if they are part of a contractual and bilateral strategy. In contrast, the legitimacy of standards in the ISO setting outside the United States, particularly in Europe, endorses the principle of national delegation. The diffusion and adoption of standards is consistent with the endogenous logic of territorial sovereignty. However, EU plans in the domain of services may lead to a dual model, in which direct participation would complement the national delegation model. This was thoroughly discussed in the consultation process preceding the adoption of the reform of the European standardisation system (Regulation 1025/2012). However, one should be aware that this would rely more upon the involvement of stakeholders within the European context than the broadening of standard recognition beyond the confines of the EU. This falls short of defining a dedicated procedure for setting future service standards. Finally, the new generation of preferential trade agreements clearly impacts on the spatial continuum along which the power of standards is likely to be recognised. Beyond the implausible prospect of new harmonised standards and the extent of mutual recognition of existing standards and regulation confined to states who take part to the agreement, we saw that parties who will not be part of the agreement will have difficulty in gaining such recognition for themselves. This applies even more in the domain of services, despite the relatively open WTO framework on mutual recognition of regulations and standards towards third parties.

* * *

This chapter has examined the authority of standards within the broader regulatory environment of capitalism by studying the case of service standards in the context of the ISO, European, and American standardisation systems, as well as the prospects of the new generation of preferential trade agreements such as CETA and CPTPP. It showed the intricate and manifold ways in which the ambiguity of the world of standards supports its power across institutional specificities. This not only goes against the view of a compelling transatlantic divide. It also calls for mitigating speculations on the prospects of current and future mega-trade agreements against the setbacks initiated by the Trump administration. With or without deals, the ambiguity on which the authority of standards feeds the regulatory environment of capitalism is here to stay. Nevertheless, over the last few years, developments in service standards have been weaker than expected. The special case of services is a first explanation. American practitioners tend to deny the distinctiveness of service standards per se, while in the European and ISO contexts on-going struggles take place to define what exactly this category may mean and why it would need dedicated procedures likely to better support the development of service standards. An alternative explanation may be that inferring a weak development of service standards reflects a fallacy of composition, as many international standards are developed elsewhere, whether or not tagged 'service-related'. This sets the agenda for examining other ambiguous and neglected aspects of the transnational hybrid authority of standards. I will begin with standards for the insurance industry, which are, as we will soon see, among the farthest from the standardisation system as usually conceived within ISO and European arenas.

5 Doubling Security

Prudential Standards for Insurance Regulation

Insurance seems incredibly boring. Each time I mention the topic, the person I am speaking to comes back to me with images of salesmen that bother you trying to sell useless and incomprehensible policies over the phone, the Internet, and door-to-door. This is certainly true sometimes and probably why Woody Allen is known for having said: 'there are worse things in life than death: have you ever spent an evening with an insurance salesman?' And yet, as Doyle and Ericson point out, 'insurance has been part of the fine print as modernity has unfolded in all its complexities, fine print that we may seldom force ourselves to read but that we ignore at our peril' (Doyle and Ericson, 2010: 244). So, before going further, here are some basic figures that will set the record of these bits of fine print straight and emphasise that we should never ignore the importance of insurance in contemporary global capitalism as well as in our daily lives.

Insurance companies and pension funds (to which the life insurance industry is closely related) account for more than half of total institutional assets under management in OECD countries, an amount estimated at more than US$50 trillion in 2013. In absolute terms, the United States owns by far the majority of insurance and pension fund assets under management in all OECD countries; the total was over US$21 trillion in 2013, and that accounted for more than half of all institutional assets under management in the country. When valued as a percentage of GDP in advanced markets, insurance companies and pension fund assets under management account for well over 100 per cent of GDP in all ten largest OECD markets except Germany, and over 200 per cent in the United Kingdom and the Netherlands (see Table 5.1).

Total expenditure on insurance premiums in 2015 is as high as 7.76 per cent of GDP in advanced markets, with an average of around $3,500 of premiums paid per capita; in emerging economies, those figures are, respectively, 3.34 per cent of GDP, with $166 of premiums per capita (Swiss Re, 2018b: 37). Although only a small portion of premiums currently originate from this latter market, at the current growth rate it is estimated that emerging economies will constitute within less than ten

Table 5.1 *Assets by institutional investors in the OECD countries: Ten highest by total amount, 2013, millions of USD*

	Investment funds	Insurance companies	Pension funds	Public Pension Reserve Funds	Other (1)	Total	Insurance and pension funds assets as % of GDP (4)
United States (2)	16,683,671	7,508,050	13,941,616	2,764,431	433,477	41,331,245	135
Japan	3,344,575	4,053,603	1,331,231	1,223,863	..	9,953,273	121
United Kingdom (2)	1,079,284	2,467,204	2,676,146	..	..	6,222,633	219
France	1,654,993	2,993,008	11,860	..	..	4,659,862	125
Germany (2,3)	1,745,806	2,295,281	235,474	..	..	4,276,561	75
Luxembourg	4,022,279	203,506	1,323	..	..	4,227,107	449
Canada (3)	1,272,459	700,690	1,260,157	277,766	..	3,511,072	134
Netherlands	628,699	586,355	1,381,901	..	..	2,596,954	265
Australia (2)	320,099	406,281	1,458,132	85,597	86,938	2,357,047	183
Switzerland (2,3)	525,875	701,033	805,462	..	..	2,032,370	358
Total OECD (2)	34,906,446	26,075,932	24,745,764	5,101,316	1,811,323	92,640,781	

Note: ".." means missing. Data in this table were used to produce figure 1 of the newsletter *Pension Markets in Focus 2014* (www.oecd.org/daf/fin/private-pensions/Pension-Markets-in-Focus-2014.pdf). Book reserves are not included. Pension funds and insurance companies' assets include assets invested in mutual funds, which may also be counted in investment funds.

1. Other forms of institutional savings include foundations and endowment funds, non-pension fund money managed by banks, private investment partnerships, and other forms of institutional investors.
2. Data are preliminary or estimated for at least one of the institutional investors (pension funds, insurance companies, investment companies, or other).
3. Data for investment companies refer to open-end companies only.
4. 2013 GDP in millions of US$ at constant prices and PPPs, OECD base 2010.

Source: OECD Global Pension Statistics, Global Insurance Statistics and Institutional Investors' Assets databases, OECD staff estimates, and my own calculus for insurance and pension fund assets as percentage of GDP.

years more than a quarter of the global insurance market.[1] Beyond market share considerations, insurance services are key market integrators closely related to the financialisation of contemporary capitalism. They lie at the core of the post-crisis accumulation regime. As banks' long-term lending shrinks and governments are set to be durably stuck in austerity gear, they have become key players in financing long-term investments such as in infrastructure, innovation, education, and health. As pension schemes and pension funds increasingly depend on complex financial products offered by life insurers and investment banks, insurance services have also gained considerable prominence in the choices that our societies face with the challenge of an ageing population. Historians consider that 'over the past three centuries, the insurance business has grown into a global colossus [... that plays] a pivotal role in redefining the contours of social solidarity, the boundaries between speculative and prudential behaviour, the basis of social and intellectual authority, the extent of property relationships, and the balance between public and private action in promoting social welfare' (Clark and Anderson, 2010: 4, 6). To what extent, then, has the insurance industry become a global colossus in the current expansion of the tertiary sector? And more specifically, how does it rely on standards to access and create markets to an ever-larger part of the world's population?

This chapter and the following one aim to respond to these questions by shining a distinct spotlight on how insurance is a significant institution of informal governance and alternative sovereignty. Both chapters show how core principles and standards are key instruments in controlling, transferring, and distributing risks in a wide range of domains which also elicit state intervention. In contrast to conventional views that standardisation and internationalisation of the service economy may be easier in a non–ideal-typical industry such as insurance that is neither immaterial nor relational, my analysis provides ample evidence that setting such standards remains, in most cases, difficult and contentious. By paying particular attention to standardisation processes, I probe the overall argument of this book on the ambiguous authority of standards that support the expansion of the tertiary sector with potentially conflicting definitions of quality and security requirements.

For greater clarity, the analysis is divided into two chapters, which focus on standards for market regulation and creation, respectively. Markets require coordination mechanisms, irrespective of policy objectives fulfilled in regulatory standards; and such coordination mechanisms

[1] Swiss Re, Economic Research & Consulting, correspondence with the author.

rest on standards. The distinction echoes the difference between coordinative and regulative standards used elsewhere for studying interoperability in technical systems (Schmidt and Werle, 1998: 120). For instance, at the turn of twentieth century, emerging industries in the field of electricity and communication such as Bell, General Electric, or Siemens developed their own interoperability standards. Likewise, at the turn of the twenty-first century, the (re)insurance industry is inventing standardised formats of data collection, exchange, and valuation. Whilst the subsequent chapter focuses on such developments of standards invented to support the creation of new markets, to reinforce existing ones, and to preside over changes in their work (in particular, in the distinct lines of natural catastrophes and life insurances), this chapter stays within the confines of the regulation of insurance markets. Following some background on the insurance industry – an industry that too often remains an obscure object of global finance and governance – this chapter shows how standards deeply affect the regulation and the supervision of the insurance industry in the post-crisis era. It then examines in some detail the most controversial provisions of the European Directive Solvency II, the most ambitious regulatory overhaul ever undertaken for insurance industries, with tremendous implications across the industry and way beyond the European Union. After that, it briefly outlines how Solvency II set the stage for developments at the global level under the aegis of the International Association of Insurance Supervision (IAIS) and regulatory policy reforms in the United States. In a nutshell, the chapter shows that the protection against risks sold by insurers obviously aims at providing security to the policy holders who buy them. Yet, policy holders may rightly ask for additional guarantees about the ability of the insurer to pay the promised sum should the insured event occur. Prudential standards exist precisely to respond to such calls for doubling the security of insurance policies. There is, however, no single way out of defining standards for such a double security. Let us see, then, what those conflicting views might look like.

Insurance: That Obscure Object of Global Finance and Governance

The service sold by insurance companies is a protection against risk paid by the insured as a defined price in what is called a premium. It takes the form of an insurance policy which, on the one hand, provides to the policy holder the contractual right to claim that protection should the insured-against event occur and, on the other hand, commits the insurance company to pay if and when such a time comes. For insurance companies, promises to pay policyholders are financial liabilities, for which they must be sure to have the money from day one to far into the future. To guarantee that protection, their task is to spread risks among the greatest and most diversified set of policy holders in order to diminish their exposure to a certain type of claim, or even a single claim too big to pay. For decades, let alone centuries, actuaries have used probability calculus to model matrixes and curves of potential losses and their frequency against which to price the premiums charged to policy holders. The weight given to the geographical distribution of potential losses, their frequency, and their size – that is where, how often, how severe the event may be – will depend on the line of insurance concerned. For instance, a large difference exists between life and natural catastrophe insurance. Having any trustworthy knowledge to estimate future losses from natural catastrophes such as earthquakes, windstorms, and floods is extremely difficult; the geographical distribution of the loss has a huge impact, with major fluctuations in size in case of extreme events (think of Fukushima!) and whose frequency is so low that there is no reliable historical data series upon which to build probabilistic calculus. In contrast, for life insurance, actuaries have built solid probabilistic and statistical knowledge to derive life expectancy estimates from mortality tables aggregating data such as age, gender, socio-economic class, smoker status, and other health-related information. In this case, the geographical distribution of the loss has less impact, the frequency is high, and the size of the loss has minor fluctuations and tends to be evenly distributed in the portfolio (risk management is like controlling for the accumulation of billions of rain drops, in contrast to a sudden flash flood[2]). All in all, the larger, the longer, and the more granular the information gathered, the better the probability calculated – and, most likely, the higher the company's profits.

This is, however, only the liability side of the balance sheet. On the asset side, an insurance company holds reserves to cover those liabilities. Those reserves are made up of various assets, such as its shares and the premiums paid by policy holders. As insurers are contractually bound to the promise to pay the insured events, even those that may occur far in the future, they face a particularly difficult trade-off between safety and long-term economic return when investing this reserve capital in the economy. As Zhang emphasises, 'there are no investments in the economy as certain and as guaranteed as promises made by insurance companies. By definition. The unavoidable implication is that insurers' assets

[2] I thank Matthieu Leimgruber for this metaphor.

can never be as securely guaranteed as their liabilities – which those assets are supposed to cover' (Zhang, 2014: chap 4). How do insurers manage the risk that their assets might lose value in the future and, thus, compromise their promise to policy holders? For a long time, the basic tool at hand has been to invest assets in low risk and long maturity instruments, such as real estate and high grade corporate and sovereign bonds, with special attention paid to the diversification of the portfolio on both the asset and liability sides of the balance sheet. Another long-established technique is reinsurance. In order to share a portion of the risks included in their portfolio, insurers use the services provided by specialised reinsurance companies that take over that part of the risk in return for a corresponding part of the premiums. This is particularly used for high loss and low frequency hazards such as natural catastrophes; but it has also been used since the 1890s in life insurance for hedging so-called substandard risks – those regarded as so high and extraordinary that they were previously insured with a hefty surcharge or, more commonly, excluded from access to a life insurance policy (Lengwiler, 2009).

While safety, diversification, and reinsurance have been used across the industry since its early days, securitisation is a more recent development. It profoundly transformed the way insurers do their job. In the same way as the banking industry has invented sophisticated instruments to pool various types of debts into securities such as the infamous collateralised debt obligations (CDOs) that gained centre stage with the global financial crisis, insurers now commonly turn insurance policies (their liabilities) into securities sold off to investors on global capital markets. Basically, securitisation is the process by which something which is not a security is converted into a security, that is, into a capital market instrument. It enables insurers to transfer risk from themselves to investors in capital markets. This involves ceding the risk to a special purpose vehicle (SPV) in charge of issuing securities and using the proceeds from the sale to pay out any claims emerging from the risk transferred (Ramella, 2010: 230). While the technique has been pioneered in the domain of natural catastrophes, it is now widely used in the arrangements used for transferring risks from pension plans and pension funds to life insurance and reinsurance. As (re)insurers are seen to have only limited capacity to accept this transfer of risks, capital market solutions are increasingly viewed as a promising option for hedging the risk that pension plans and annuity providers are not willing or able to retain via a capital buffer. Recent developments in the securitisation of the life insurance industry thus give rise to much overlap with the pension and finance industries. It brings the industry ever closer to investment banking and shapes new demands for pricing and regulatory standards.

Insurance services thus control, transfer, and distribute risks in a wide range of domains in which states can intervene as well. More importantly, by pooling risks into sophisticated actuarial tools, insurance products sold by companies shape multiple and contradictory forms of private governance beyond state control at an increasingly global level. According to Ericson, Doyle, and Barry (2003: 14), "insurance is even THE main institution of governance after the State". Differing from one country to another, certain lines of private insurance are mandatory, such as those for cars, occupational accident, or conversion of pension annuities. In other cases, they are not, but can be compelled upon request from one party to a contract (for renting an object, for instance). General conditions, information provided, exclusion clauses, and so on confer to insurers a role of 'extra-legal regulators'. As Heimer (2002: 128) points out, 'in requiring insurance coverage as a condition for operating a business, owning a home, driving a car, holding office, or engaging in any number of activities, governments, employers, banks and other organisations are also requiring policyholders to follow insurers' rules'.

In the broadest sense, then, the insurance industry looks like an institution of informal governance resting on a system that, although largely behind the scenes, remains closely connected to state power in its capacity to exert control at distance in counterpart to security guaranties.[3] Very few studies have investigated the pioneering hypotheses of the late Susan Strange and Virginia Haufler on the ambiguous authority of the public/private nexus of insurance services across domestic and global realms (Strange, 1996: 122–134; Haufler, 1997). Strange emphasised that to understand how 'more and more lives and fortunes are affected by the ways in which ... the insurance business is conducted', we should return to the key questions of how does it exercise, '"power over what and whom? And in the end, in whose favour does it operate, and at whose cost?' (Strange, 1996: 123 and 124). Power issues in the insurance industry are clearly all-pervasive, but what matters here is that to give effect to such forms of private authority on an international plane, insurance companies – as in any other industry – set guidelines and standards reducing uncertainty in the delivery and consumption of their services. From an institutionalist approach à la North, Haufler (1997: 16) underlines that insurances are basically nothing else than intermediations in market transactions, transforming uncertainty into risks sufficiently measurable to objectify the required conditions to provide against them. From this perspective, standards provide the best way to influence the provision of

[3] See Chapter 1 for further references to studies in sociology and history.

public as well as private security against risk: 'the relevant norms and practices concern both the treatment of customers and the ways insurers, reinsurers, and ancillary businesses cooperate among themselves'. In political risks studied by Haufler, insurance principles and standards play a key role in framing the behaviour of market actors, especially in the domain of export credit and investment guarantees and marine insurance. Lobo-Guerrero's inspiring trilogy on insurance gives more emphasise to substantive and normative issues (Lobo-Guerrero, 2011, 2012, 2016). By combining Foucauldian approaches, security studies, and international political economy, insurance is understood as a technology of government promoting and protecting distinct lifestyles. From this perspective, insurance transforms uncertainty into risk through a complex process by which it 'renders uncertainty fungible' (Lobo-Guerrero, 2011: 4). This goes back to the classical age, with the invention of life insurance policies and the strengthening of maritime insurance. In the present day, this form of power exercised by insurance can be found in domains as diverse as environmental risks, kidnapping, or health insurance. According to Lobo-Guerrero, the strings of this peculiar form of 'insurantial sovereignty' reconstitute the international, all the more with the recent development of liberal governance practices 'premised on the capacity to transform uncertainty into risk and to act upon it through risk management partnerships and schemes' (Lobo-Guerrero, 2012: 125). This has recently been reinforced by the growing use of capital markets to complement old-style actuarial calculus for hedging risk portfolios. Together with highly sophisticated simulation and modelling techniques, the securitisation of life insurance is thus seen as a strategy to 'liberate insurability from the temporal strictures of traditional actuarial practices and create an infinite space for market development' (Lobo-Guerrero, 2014: 366). Securitisation, simulation, and risk modelling unmistakably support an insurance industry that brings together powerful transnational forces shaping a global finance-led accumulation regime. Yet, the financial manoeuvres, mathematical calculus, and asset management techniques used by insurers and investment bankers to issue life-related bonds need additional qualification against some agreed benchmark before finding a swift pathway on capital markets. Otherwise, the market would never be liquid enough to offer any prospects of 'infinite space for market development'. In other words, standardisation is part and parcel of securitisation.

Those few studies take due account of norms of behaviour and institutional forms upon which private insurance contracts rely to provide security on a scale that transcends states' territorial sovereignty. Yet, in contrast to accounts of insurance governance in terms of discursive regimes, governmental rationalities, securitisation, and modelling strategies, this chapter and the following one focus on how the insurance industry relies on standards to control, transfer, and distribute risks as well as to avoid, as much as possible, state intervention. In paying particular attention to standardisation processes, I continue my journey on the ambiguous authority of standards that support the expansion of the tertiary sector. Let us then see to what extent the insurance industry uses standards to provide guarantees against opposing understanding of quality and security uncertainties. I begin with the post-crisis supervisory and regulatory environment.

Supervising and Regulating Insurance after the Crisis

Apart from the special case of AIG, who benefited from the largest bail-out in the history of the United States after having sold too many securitised products with too few guarantees to Lehman Brothers, insurance companies were not at the crux of the global financial crisis. Nevertheless, the volume of their assets under management, the growing convergence of the industry with banking and other financial services, and the subsequent systemic risk borne by the largest companies have spawned considerable efforts to bring insurance regulation in line with the more stringent rules directed to finance. According to industry experts, officials of regulatory agencies, and analysts, the reforms of the regulatory environment in the post-crisis era brought a revolutionary change to the insurance industry, with a major expansion of public oversight and tighter definition of industry self-regulation (Monkiewicz, 2013). The package brought together with the Solvency II Directive of the European Union undoubtedly assumes a leading role in this regard. This is why our analysis focuses almost entirely on the standards set in motion by the European regulatory environment, with only a cursory examination of what has happened in a direct line from the Solvency II framework in the United States and at the multilateral level. As Gideon Benary, founder and editor of the boutique online publication *Solvency II Wire*, makes it plain, 'the whole drive for global insurance regulation comes from Europe; the EU clearly drives the pack'.[4] Similarly, JP Morgan, arguably the investment bank with the closest links to asset management and specialised financial products for insurers on the London market, defined Solvency II as a 'game changer' that made the power of regulators full and comprehensive.[5] Be that as it may, the

[4] Interview with Gideon Benari, Editor of *Solvency II Wire*, London, 31 March 2015.

[5] JP Morgan, 'European Insurance – Solvency II: a Potential Game Changer.' *Europe Equity Research*, 19 January 2010 (quoted in: Hall and Berset (2010: 97)). According to

following analysis shows that the change in the game made by Solvency II is as much about the power of the regulators as it is about conferring authority on standards and internal models in which the industry as a whole – and the biggest firms in particular – have considerable clout. Yet, before going into some details of Solvency II, let us have a brief account of why standards are in a position to stand as regulation in the shift towards risk-based regulation that has taken place over the last two decades.

Risk-based regulation has gained momentum following the rise of financialisation and globalisation. By and large, it substitutes for rule-based regulation considered too costly and too inflexible for keeping pace with market innovation and timely supervisory intervention. Far from mere deregulation, it sets general principles and countless technical criteria devised to foster market-based incentives and the use of business self-regulation tools. In the field of insurance – as in banking and elsewhere – the argument goes, each company's products and liabilities are unique in detail and the global and highly competitive marketplace makes flexibility and responsiveness indispensable in an ever-changing environment. Zhang reminds us, not without irony, that risk-based approaches reflect a broader principle-based regulation in which, 'fixed rules and standards are impossible to define – let alone to meet or enforce. The only body with the necessary data and knowledge to evaluate the operations and status of a 21st-century insurance company is … the insurance company itself' (Zhang, 2014: chap. 3). While the rationale of regulation in a rule-based system is legalistic and tends to take the general public as the prime reference point, risk-based regulation takes an economic approach, bringing on board other stakeholders, such as management, shareholders, and market players to whom the regulator is supposed to respond.[6]

Paul Fisher, Executive Director at the Bank of England and former Deputy Head of the Prudential Regulation Authority, Solvency II is much more rule-based than the Individual Capital Adequacy Standards (ICAS) used so far in the United Kingdom with more leeway for the judgement of both regulators and regulatees (interview with the author, London, 28 April 2015).

6 While using the economic approach, regulatory agencies are now furthermore split between two different models of the consumer: the first follows a neoclassical understanding of free market rationality; the second uses recent approaches in behavioural economics. Behavioural regulation considers that the consumer lacks the cognitive capabilities and financial literacy supposed by the neoclassical approach. Therefore, the regulation needs to 'nudge' the consumer to improve her/his supposed judgment, were s/he fully informed or well advised. Cass Sunstein, who coauthored the book that made behavioural regulation famous (Sunstein and Thaler, 2008), was the Administrator of the White House Office of Information and Regulatory Affairs in the first Obama administration. According to David Blake, founder and director of the

Risk-based regulation confers authority on standards in two distinct ways. The first remains within the confines of the technical authority of regulatory state agencies. In the post-crisis environment, their authority has indeed considerably increased with their ability to devise standard formulae for implementing general principles in such a way as to remain responsive to the numerous and highly complex dimensions in which market transactions take place. This leaves only a small circle of accomplished professionals in actuarial calculus, risk management, accounting, and neighbouring fields of finance and economics in a position to draft technical specifications wherever they are not left in the hands of modellers hired by large companies authorised to set their own internal models. The second way in which risk-based regulation confers authority on standards sanctions the expertise of private companies that can, in this new framework, apply their own internal models as an alternative to the standard formula designed by the public regulator. The following example provides a shining illustration of the ability of risk managers to turn internal models to their own advantage. In the United States, so-called variable annuities have been among the most profitable products sold by life insurers for many years. They provide a kind of financial insurance policy with a guaranteed floor for the policyholder's investment whatever happens on the markets. In normal times, policyholders will not make claims. Their funds will face short-term variations, yet remain above the floor value guaranteed by the policy. In times of a widespread downturn, however, the guarantees are likely to kick in and generate a systemic contagion that is very difficult to diversify away. A proper calculation of the reserve and solvency capital required to cover this risk is therefore critical (the AG43 guidelines of the National Association of Insurance Commissioners (NAIC)). It can be done either on a standard scenario with a set of stipulated assumptions prescribed by the regulator or on the insurer's own projections and self-selected set of assumptions (i.e. an internal model). In a preliminary study for the implementation of the guidelines, an insurer was able to tune the self-selected assumptions in such a way as to lower its internal model reserves considerably below the standard scenario reserve. The strategy to minimise the requested reserves amounted to more than half of the original standard scenario reserve designed by the regulator.[7]

Cass Business School Pension Institute, 'we can't ignore behavioural economics; the Bank of England now has a behavioural finance group, like many other regulators and governments' (interview with the author, London, 20 April 2015).

7 The example is borrowed from Zhang (2014: part II).

This example clearly highlights that the shift towards risk-based regulation not only lends credence to prospects of capture by industry interests and opens space for the ambiguous authority of rules and standards almost impossible to define. It also reveals the democratic deficit of the increasing power of the regulatory state. This is particularly true with regard to the intricacies of the decision-making and oversight procedures of the European Union and the jungle of the federal and state-level regulations in the United States. As we will see in further detail, the European Insurance and Occupational Pensions Authority (EIOPA), the regulatory body in charge of implementing rules and technical standards, fought hard to impose its view on the new regulatory power sanctioned by Solvency II. However, it backed down on a number of key issues that would have forced companies to increase their capital buffers.

The Astonishing Power of Solvency II

Solvency II is the system for insurance regulation set up by the European Union and implemented as of 2016. Its main objective is to strengthen the insurance regulatory regime with a set of new EU-wide harmonised capital adequacy, risk management, and reporting requirements for insurers. It has been established with a view to reducing the prospect of bankruptcy and market disruption in insurance and therefore reinforcing the ultimate protection of policyholders. The Solvency II Framework Directive (Directive 2009/138/EC) was adopted in 2009 and amended in 2014 after strenuous negotiations (Directive 2014/51/EU, the so-called Omnibus II Directive). Following the Lamfalussy process used in the European Union for the regulation of the financial service industry, the framework directive (level 1) is a full legislative process involving the Council and the European Parliament; it sets out general principles, whose implementation is defined more precisely at lower levels. Whilst the European Commission and regulatory agencies consult each other for the definition of the technical implementing rules setting out the so-called delegated regulations (level 2), regulatory agencies are on their own in defining technical standards and guidelines of the so-called levels 2.5 and 3. The Delegated Regulation (EU) 2015/35 on Solvency II and Omnibus II was published in January 2015 with 381 articles (and 797 pages with annexes) of binding rules, directly applicable to all member States of the European Union and the European Economic Area. In 2015 the European Insurance and Occupational Pensions Authority (EIOPA) furthermore published sets of Implementing Technical Standards (ITS) as regulatory tools subsequently endorsed by the Commission.

Like the Basel framework for banks, the risk-based regulation of Solvency II rests on a comprehensive system of quality and security standards divided into three pillars. Pillar 1 addresses quantitative requirements, with sophisticated technical specifications for the valuation of assets and liabilities used in the calculation of capital requirements. Pillar 2 introduces qualitative requirements related to the responsibility of the insurers themselves to manage their risks and governance structure; it includes in particular a new enterprise risk management system called Own Risk and Solvency Assessment (ORSA). Pillar 3 sets out rules on transparency, reporting, and public disclosure with the objective of enhancing market discipline, information, and competition across national jurisdiction.

The most significant change brought by Solvency II is a calculation of capital requirements based on a 'total balance sheet approach' that aims to better take into account the specific risks borne by each insurer and, accordingly, the amount of capital it should keep in reserve to hedge those risks. Capital requirements are defined along a two-step ladder. First, the solvency capital requirement (SCR) sets a level of resources that enables the absorption of significant losses and gives reasonable assurance to policy holders and beneficiaries that payments will be made as they fall due.[8] Then, an additional minimum capital requirement (MCR) sets the lowest level of capital below which the resources should not fall, failing which an intervention of supervisory bodies is triggered and may go as far as a withdrawal of the authorisation. The thousands of items included in the valuation of assets and liabilities must follow a market-consistent approach. To this end, they use, wherever readily available, a mark-to-market approach already used in other regulatory frameworks on banking (Basel II & III) and accounting (IFRS). However, in many cases this is not possible, and alternate principles, guidelines, standard formula, as well as criteria that allow for large companies to set and use their own internal models are required.[9]

This is where the authority of standards in the regulatory environment set by Solvency II begins. Examining the several hundreds of modules and sub-modules of risk defined by Solvency II is beyond the scope of

[8] The 99.5 per cent confidence level set by Solvency II is equivalent to calculations based on a risk of loss caused by a 1 in 200-year event.

[9] According to research conducted by *Solvency II Wire* ('Delivering Solvency II internal models', 4 May 2015, online), approximately 175 insurance and reinsurance entities across Europe were in pre-application for an internal model or a partial internal model in late 2014 in preparation for the implementation of Solvency II in 2016. EIOPA, the European regulator, used a slightly different methodology, based on insurance groups, not individual entities; the figures were therefore smaller, with just over 100 pre-application processes under way in late 2014.

the present analysis. I choose instead to point out the ambiguous authority of standards in the most controversial issues related to the implementation of Solvency II.

The successive delays and the high level of technical detail included in the Directives themselves provide primary evidence of the power of standards in Solvency II and Omnibus II. Following the 2002 Sharma Report that initiated the wide-ranging Solvency II reforms, the first timeline set an implementation deadline for 2008, subsequently delayed to 2012, and then again to 2014. In 2011, further delay took place after the shockwave sent by the quantitative impact study (QIS5) published by EIOPA, which showed that in the aftermath of the global financial crisis the application of Solvency II standards would put many companies in a much tougher situation than expected. Negotiations almost froze. According to Gideon Benari, founder and editor of *Solvency II Wire*, 'the industry completely panicked and put all its force against the European Parliament and Council'.[10] Accordingly, the numbers had to be changed with new ways of devising many items included for calculating the solvency capital requirement; moreover, the industry could not leave those detailed measures in the sole hands of a regulator who so blatantly harmed its profit expectation. Strenuous negotiations followed, during which the latest deadline was eventually fixed for January 2016, with a transitional period of twenty (sic) years obtained after heavy lobbying by the German insurance industry, which considered itself as the most affected by the change introduced with the mark-to-market model. In the words of an insider, Napoleonic wars were briefer (Smolinski, 2013)! Once industry lobbies won over the European Parliament and Council that the market consistent valuation of all assets and liabilities would have dramatic effects, they furthermore sought direct legislative control on technical standards on all related matters. While Omnibus II was at first only designed to amend the Solvency II Directive in accordance with the establishment of EIOPA and its new powers in coherence with other post-crisis regulatory agencies, it eventually included many technical specifications supposed to be dealt with at lower levels of the Lamfalussy process. The aim of the industry was indeed to ring-fence at the highest level as many standards as possible. There is no need to point out that this not only strengthens the balance sheet of insurers but also threatens the protection of policy holders and reinforces the likelihood of using taxpayers' money to bail out insolvent companies.

The so-called Long Term Guarantees are the most controversial issue of Omnibus II. They provide a set of detailed measures designed to

[10] Interview with Gideon Benari, Editor of *Solvency II Wire*, London, 31 March 2015.

adjust (and discount) the calculation of solvency capital requirements by taking into account the significant share of long-term liabilities held by life insurers that in principle are not subject to the same risk of market volatility as for other actors of financial markets. But to what extent? The pervasive presence of standards when it comes to responding in concrete ways to the question gives us a measure of their transnational hybrid authority. In the discussion that follows, I show how the defining criteria conferring authority to those new cross-border regulatory practices rest on much ambiguity. Depending on the precise way they are designed, these technical specifications have huge implications in terms of the three-dimensional framework of the private/public institutional continuum, the technical/societal material continuum, and the national/transnational spatial continuum on which the power of standards is situated. I shall come back to this after having presented those technical provisions in some details.

Annuities and many other life insurance products related to savings, pensions, and retirement offer long-term guarantees to policyholders. Some life-related annuities sold by German insurers, for instance, have pay-outs that fall due in more than fifty years' time. Obviously, an insurer holds reserve assets to cover the value of its liabilities. But how to value such liabilities in a mark-to-market total balance sheet approach is a much trickier question – almost impossible to calculate indeed! The Omnibus II Directive introduced so-called long-term guarantee measures to address the technical details according to which the rules for the market-consistent valuation of assets and liabilities should be set. Basically, the measures address three adjustment mechanisms to calculate capital requirements in a way that is supposed to balance the interests of policy holders' protection and of insurers' to hold less reserve and therefore have more assets freed up for other purposes.

The first measure aims at calibrating the capital requirement to a 'matching adjustment'. Even if insurers can never exactly match long-term liabilities with corresponding assets, their balance sheet is not exposed in the same way as banks may be to short-term volatility in the valuation of assets and interest rates. For such assets are in principle held to maturity and the much longer term of the liabilities prevents any risk of forced sale of the assigned portfolio of assets. It is against this background that the measure applies so-called discount rates whenever insurers can match predictable long-term liabilities with a replicating portfolio of assets. A basic job for insurers is to set up provisions for the future payments to policyholders. A trickier one is to apply discount rates to calculate the value at which each of their liabilities can be reduced as a function of the average interest rate expected for the duration of their maturity. The value of a liability today can thus be discounted at an amount that will depend of the level of future interest rates for the whole time that liability stays in the balance sheet. As life insurance policies sell promises to pay out annuities many years after the policy is contracted, small differences in rates applied for the discount can make huge differences in valued amounts. The higher the discount rate, the lower the value of the liability in today's money. Inversely, a small discount rate reflects a little valorisation expected in the future and thus limits the ability to lower the value of the liability in today's money. Unsurprisingly, insurers are big fans and strong advocates of high discount rates that let them free up more available capital.

This might look like a mere battle of numbers, with EIOPA initially backing much higher figures than the industry. Yet, this hides clear and present danger for pensions backed by life insurance. For instance, according to the methodology derived from the Solvency II regime and tested by a quantitative impact study (QIS) of EIOPA, Sweden's workplace-based defined benefit schemes typically boasted a surplus of assets over liabilities of 13 per cent. By contrast, there would be a deficit of 24 per cent in the United Kingdom compared to existing rules, and even a deficit of between 81 and 93 per cent in Ireland. As highlighted by Jane Beverley, head of research at Punter Southall, a consultancy, 'the results demonstrate that the impact of applying a Solvency II-style regime to pensions could be huge'.[11] The prudent and low discount rate sought by EIOPA clearly limits market expectations for actors operating in the most market-driven insurance and pension funds environments such as Ireland, the Netherlands, Finland, and the United Kingdom. But at the same time, EIOPA's strong regulatory posture aims to strengthen the unification of the European insurance and pension market space.

Actually, member States could already take into account the effects of long-term asset-liability management strategies in valuing their insurance liabilities and the corresponding assigned assets backing them.[12] The issue at stake was to reach a harmonised standard against divergent implementation techniques used in each jurisdiction. Aside from debates related to fundamental flaws of the measure itself, the scope of assets and liabilities to be included in the calculation has been a stumbling block right from the start.[13] Above all, the German life insurance industry, holding a large book of liabilities with long-term guarantees, pushed for the broadest scope, in opposition to the much more prudent accounting practice of UK annuity

[11] 'EU pension proposals alarm local schemes', *Financial Times*, 21 April 2013.
[12] Article 20.1.B.(a)(ii) of Life Directive 2002/83/EC.
[13] For an accessible technical presentation, see in particular: Danielsson et al. (2012).

writers, with whom the whole agenda began. Despite countless attempts to reach precise definitions at the highest institutional level, the Directive does not fully define a closed list of admissible assets. Instead, it only defines certain behavioural features of the entire asset portfolio likely to be eligible. Accordingly, standards setting the capital relief for a 'matching adjustment' of assets and liabilities remain uncertain and ambiguous.[14] Both insurance companies and national supervisory authorities will compete to impose their view according to the interests of the industry, without the larger public of policy holders, who are supposed to be better protected, having much a say.

The second technical measure conferring considerable power to standards and how to define them is a "volatility adjustment". Here, the objective is to increase the discount rate to avoid artificial depreciation of the balance sheet in times of stressed economic conditions. For the Commission, this wider application of discounting liabilities should help to avoid pro-cyclical investment behaviour of insurers when bond prices deteriorate owing to low liquidity of bond markets or exceptional expansion of credit spreads; the adjustment thus aims at stabilising the capital resources of insurers in times of crisis rather than the other way round.[15] Like all aspects of the long-term guarantee measures, finding common ground for calculation of this volatility adjustment has been plagued with technical issues, let alone profound doubts on an adjustment measure whose lack of economic foundations makes it easily amenable to capture (Danielsson et al., 2012). Beyond this, the measure adjusts reserve requirements only in distress, contrary to a fully countercyclical buffer approach that would provide for additional reserve accumulation in boom time. Such a lack of symmetry in the adjustment makes it look like a one-way pendulum: it kicks in to the advantage of insurers when markets fall below normal, but does not turn against them with further shock absorbers when they rise above normal (Zhang, 2014: chap. 19). According to Francesco Mazzaferro, Head of Secretariat of the European Systemic Risk Board, and his colleague Jeroen Brinkhoff, this lack of symmetry 'creates an incentive towards risky behaviour [and] gives regulatory relief to insurers' (Mazzaferro and Brinkhoff, 2012). While the standard was first conceived to take due account of the fact that insurers

[14] 'Solvency II uncertainty. A reality', *Solvency II Wire*, 13 October 2014.

[15] European Commission, Solvency II Overview – Frequently Asked Questions, 12 January 2015. This proposal was first advocated by French insurers, who invest heavily in equities and would therefore be more affected than others by profound distress in markets; insurers from Italy, Spain, and Portugal followed suit at a time when they held a lot of bad sovereign debt in the middle of the euro sovereign crisis; for further details, see Smolinski (2013).

are in principle not affected in the same ways as others by volatile markets, it eventually leads to an ambiguous adjustment system that leaves considerable ground to protect the interests of those insurers who are precisely the most affected by volatile markets. With volatility adjustments in times of crisis, a new standard is born, less to reduce the risk borne by insurers than to ease their balance sheet.

The third aspect by which long-term guarantee measures confer power to standards is a new system of valuing liabilities, called extrapolation. It follows the move away from the market consistency approach of Solvency II already seen in the two previous adjustment measures. Here, the alleged objective is to respond to the exacerbating pressure brought on solvency positions by extra-low interest rates and the difficulties raised by the valuation of the very long-term liabilities where no assets with similar maturities exist. As mentioned earlier, German insurers, in particular, sell annuity products whose pay-outs are not due for a very long time. In principle, no risk free assets are on the market with such long maturities. Extrapolation is therefore made up to estimate the interest rate where no reliable market data exist. Instead of using mark-to-market, the asset-liability management here works on a specially calibrated basis, called mark-to-model. According to European officials, this further application of discounting liabilities by extrapolation aims to 'ensure that the valuation of technical provisions and the solvency positions of insurers are not heavily distorted by strong fluctuations in the short-term interest rate'.[16] A further advantage, however, is that extrapolation gives higher and more stable long-term interest rates for valuing the long-dated liabilities of insurers and pension schemes (Evans et al., 2013). A sticking point for raising the rates has therefore been how quickly the extrapolation of interest rates should start and the evolving yield curve converge and reach its highest level. The quicker this happens, the higher the rates – and the lower the reserves to be held by insurers. The controversial fifth quantitative impact study of EIOPA (QIS5) set the starting point at thirty years with a convergence time of over forty years. In contrast, the Omnibus II Directive sets a quicker starting point at twenty years.[17] Here again, the German insurance industry successfully lobbied to set new rates right into the Directive and increase the extrapolated rates quite substantially before the (quite optimistic) 4.2 per cent convergence rate set by EIOPA.

[16] European Commission, Solvency II Overview – Frequently Asked Questions, 12 January 2015.

[17] Directive 2014/51/EU, paragraph 30 of the Preamble.

Beyond these three distinct long-term guarantee measures, the Delegated Regulation of the Commission and the lower level implementing technical standards (ITS) of EIOPA considerably lowered the calibrations of most risk factors in the standard formula used to calculate the solvency capital requirement. Without going into the details of the countless items included in the modules and sub-modules, suffice it here to take the following example. In February 2015 EIOPA set the discount rate for liabilities in euros with a maturity of thirty years at 1.86 per cent, a figure significantly higher than the 1.48 per cent set by the European Central Bank for similar assets (Euro area yield curve for AAA rated government bonds as of 31 December 2014, maturity thirty years). Thanks to such higher figures, liabilities can be discounted to a greater extent and insurers will not need to put as much money aside to cover them. According to Sven Giegold, a Green MEP and one of the most vocal critics of Solvency II, this amounts to nothing else than a financial scandal: it 'puts at risk the security of long-term insurance, in particular of annuity insurance … EIOPA disguises the financial problems of insurance companies instead of ensuring transparency [and] violates its mandate which foresees explicitly consumer protection.'[18] Very few elected politicians are as well aware of the sweeping power issues veiled under the arcane technicalities of insurance standards. Yet, as we have just seen, there is ample room for politicising such technical specifications.

Earlier in this chapter, I stressed that Solvency II set the stage for regulatory overhauls elsewhere, and this is why I decided to concentrate on this framework in my assessment of the power of standards in insurance regulation. It is therefore beyond the scope of this book to make a complete overview of standards used elsewhere, how they replicate what we have just seen with Solvency II, and the extent to which they might on the contrary diverge from it. However, for the sake of avoiding an excessively Eurocentric view, I now briefly examine developments underway at the global level and how both Solvency II and those global developments impact on regulatory policy reforms in the United States.

A Basel for Insurers

At the global level, the International Association of Insurance Supervisors (IAIS) has also undertaken an ambitious programme of enhanced standards of global insurance supervision and cooperation. Analyses

[18] Sven Giegold, Press release, 28 February 2015, available at: www.sven-giegold.de/2015/financial-scandal-over-the-weekend-life-insurance-regulation-denies-reality (accessed 25 June 2015).

often refer to the Swiss town of Basel as a metonymy for the higher standards developed by the Basel Committee on Banking Supervision under the aegis of the Bank for International Settlement (BIS). However, as the BIS also shelters the IAIS Secretariat under its roof, the comprehensive and global framework of insurance regulation and supervision looks a lot like a new Basel for insurers. The architecture of the regulatory and supervisory requirements rests on three tiers. The first sets insurance core principles (ICPs) to be used by distinct legal entities as well as at the higher level of insurance groups. The second tier targets more specifically internationally active insurance groups (IAIGs) and establishes a Common Framework (ComFrame) of global regulatory standards specifically directed towards the fifty or so insurance groups concerned. While this common framework is built and expands upon the insurance core principles whose first version was adopted in 2011, it will set out a comprehensive range of qualitative and quantitative requirements, including a new risk-based global insurance capital standard (ICS), whose version fit to implementation by supervisors is expected to be adopted by the end of 2019. The third tier is even more closely focused, as it is designed for the ten or so global systemically important insurers (G-SIIs) identified under the purview of the Financial Stability Board (FSB) and G20 and for which higher loss absorbency (HLA) requirements are intended to address their systemic importance in the global financial system. As a foundation for those additional requirements to be applied from 2019, the IAIS developed basic capital requirements (BCR) for all group activities, including non-insurance ones. Upon implementation, global systemically important insurers are expected to hold regulatory capital that is not less than the sum of the required capital amounts from the basic capital requirements and higher loss absorbency requirements (regulatory capital ≥ BCR + HLA).

The impact of the bold framework of regulation and standards brought together by the IAIS and Solvency II joins forces with the globalisation and financialisation of insurance services as main drivers of policy reforms currently underway in the United States. A key characteristic of the US regulatory environment, which dates back to the nineteenth century, is that the insurance industry is still largely regulated at the level of the fifty states of the Union, above which the National Association of Insurance Commissioners (NAIC) adopts model laws and standards expected to be implemented at the lower level. NAIC has no power to impose them directly, but uses its accreditation authority as a strong implementation incentive. This is clearly a major hindrance for market access across states, let alone from an international perspective. The adoption of the Dodd-Frank Act in 2010 and other post-crisis reforms that attempt to match IAIS insurance core principles somewhat changed the landscape.[19] The extension of the Federal Reserve Board's responsibilities to cover consolidated supervision of insurance groups has strengthened their supervision and covers around 30 per cent of total premium income in the United States. Moreover, although insurance will continue to be regulated by the states, the Federal Insurance Office (FIO) created within the US Department of the Treasury now has greater monitoring and intervention power, in particular when state laws are considered as inconsistent with a negotiated international agreement and discriminate against non-US insurers.

One of the most contentious issues is the additional collateral requirement imposed on non-US reinsurers in more than half of the American states and the limited number of jurisdictions qualified for lower collateral and inter-state business despite a new NAIC model law adopted in 2011. With removal of such requirements within sixty months, this is one of the key issues of the US–EU covered agreement on insurance signed in 2017.[20] Another important characteristic of the US regulatory environment is its largely principles-based nature. In the same vein as qualitative requirements of the Pillar 2 of Solvency II and the IAIS Insurance Core Principle 16, increased emphasis is being placed on risk management through the introduction from 2015 of an Own Risk and Solvency Assessment (ORSA) regime. However, in contrast to the risk-based approach adopted by Solvency II with detailed and ambitious quantitative and qualitative standards, this principles-based approach provides fewer of these requirements. Instead, the regulator defines guidelines and avoids being too specific in defining the key elements of the calculation. As we have already seen, this also allows insurers to develop their own internal models, whose underlying assumptions leave ample room of manoeuvre for opposing and ambiguous interpretations. According to Benjamin Lawsky, the New York state financial services superintendent, 'companies will take every advantage of [principles-based regulation] to reduce their reserves as much as possible, [which] leaves insurance regulators vulnerable to the charge that we are too willing to sacrifice

[19] For further detail, see: International Monetary Fund (2015).

[20] Interview with Miroslaw Galar, European Commission, DG Trade, Service Unit, Brussels, 28 May 2015; Office of the United States Trade Representative, 'Joint Statement on U.S.-EU Negotiations for a Bilateral Agreement on Insurance and Reinsurance Measures', 13 January 2017; 'Bilateral Agreement between the European Union and the United States of America on Prudential Measures Regarding Insurance and Reinsurance', 22 September 2017, online at: https://ec.europa.eu/info/system/files/170113-us-eu-agreement_en.pdf, accessed 8 August 2018; for background, see Insurance Information Institute, Regulation Modernization, April 2015, online at: www.iii.org/issue-update/regulation-modernization, accessed 29 June 2015.

solvency and consumer protections in our regulation of the industry'.[21] Finally, while Solvency II and IAIS insurance core principles and future insurance capital standards apply at the group level, another key characteristic of the United States is the lack of uniform capital requirement at the level of the insurance group. Several bodies have recently been developing proposals aimed at meeting the standard set by the IAIS. Yet, according to industry experts, group capital requirements are perhaps the most difficult challenge for the US regulatory system to address. So far, consulted parties strongly support the principle of keeping the US Generally Accepted Accounting Principles (US GAAP) approach, which differs from the IAIS's market-adjusted valuation approach (KPMG, 2015: 51). Although IAIS keeps repeating that standards will ultimately be globally harmonised, a double valuation system looks like having a bright future.

* * *

It is pointless to delve further into the colossal technicalities of all the standards concerned by those major reforms of the insurance regulatory environment to reflect upon the power they exercise. While insurance as such is a prominent instance of global finance and governance, standards provide additional guarantees that the many ways by which insurers transform uncertainty into measurable and fungible risk are secure and backed by solvent companies. In contrast to views focused on the ingrained power of either public regulation or private securitisation, the argument put forward in this chapter is that it is neither strictly one nor exclusively the other. By doubling the security sold by insurers to policy holders, the power of standards as regulation results from their ability to bring together the private and public dimensions of broadly defined security and quality concerns. Moreover, in doing so, they reflect both the physical objectification of many different lines of specialised risk studied by actuaries and other insurance experts and the societal values affected by such technical specifications. Finally, the opposing political economy objectives accommodated in those regulatory standards spread out across territorial states thanks to the intertwined logic of endogenous recognition akin to sovereignty and exogenous adoption akin to market power. Ambiguous transfers of authority thus pervade the three private–public, technical–societal, and national–transnational dimensions of our analytical framework. I now turn to these dimensions, before continuing in the next chapter our journey in the world of insurance standards

[21] Quoted in Zhang (2014: chap. 3).

away from market regulation towards market creation – those unchartered routes that support the creation and consolidation of new insurance markets.

First, concerning the private and public spheres increasingly blurred in this new regulatory framework, there is no question that the aforementioned standards are all set by public regulators. This goes without saying for regulatory bodies such as EIOPA in the European Union and NAIC in the United States. Similarly, IAIS is a public body of insurance regulators and supervisors of more than 200 jurisdictions in nearly 140 countries. Nevertheless, those public actors are not immune from regulatory capture and all follow suit on principles-based regulation that places great emphasis on the ability of private companies to set their own standards. Regarding regulatory capture, we have seen particularly clearly in the context of Solvency II how EIOPA backpedalled on a number of technical specifications and calibration choices faced with a barrage of criticism from all parts of the insurance industry, in particular powerful German life insurers, the London market, and French advocates of 'bancassurance. One can always say that this is common politics. More interesting for my argument is the ability of private companies to set their own standards. Besides Own Risk and Solvency Assessment (ORSA) regimes briefly referred to in this chapter, it is worth mentioning here that IAIS Insurance Core Principle 17 on capital adequacy sets a number of criteria for the use of internal models to determine an insurer's regulatory capital requirements. In doing so, it incorporates a number of provisions earlier discussed within the Solvency II framework intending to revamp and level the playing-field of public regulation. It also sanctions the very rationale of industry self-regulation by internal models: 'where the supervisor allows a range of standardised and more tailored approaches for regulatory Capital purposes, including internal models, an insurer should have a choice as to which approach it adopts' (IAIS ICP guidance 17.12.3). In designing the subsequent ICP Guidance 17.12.4 that considers 'cherry-picking' between those approaches as inappropriate behaviour, the IAIS most probably knows only too well that this can easily be used with the explicit purpose of lowering capital requirements as compared to the standard formula set by the supervisor. The shift towards principles-based regulation, the use of internal models of solvency capital requirement, and qualitative requirements such as the Own Risk and Solvency Assessment (ORSA) put private insurers in a position to set standards on their own and shape regulation for their own favour across borders. In so doing, governmental and inter-governmental regulatory bodies support and fully recognise the self-regulatory power of private insurers. This suggests that the private or public status of standards in the enhanced regulatory environment of the insurance industry in the post-crisis era remains highly ambiguous. Indeed, public regulators tend to relinquish their responsibilities in a number of domains and hand it over directly to the companies they regulate. At the same time, we cannot deny that major developments of public regulation have taken place, with concrete and across-the-board outcomes for standards included in hundreds of modules, sub-modules, and items to follow and report on.

Second, on the material dimension merging technical specifications and societal values, those complex regulatory regimes in progress indubitably hide behind their cloak of protection bold societal values, such as greater policyholder protection and enhanced confidence in fair, safe, and stable insurance markets supporting a more secure daily life.[22] While the industry identifies consumer protection as a prominent trend in current and future regulatory reforms, its effectiveness remains questionable. The end-consumer who holds insurance policies is largely excluded from this highly technical framework. As Dough Taylor, member of the UK Financial Services Consumer Panel, points out, 'consumer representation in the financial and insurance industry remains extremely weak as it needs very high skills similar to those used in the industry and can only offer very low wages in no proportion to those practiced in the industry'.[23] Moreover, the maths used for formulas and the numbers used for calibration tend to conceal blunt truths of major social implications for the consumer behind the veil of sophisticated science. For instance, we saw the huge impact that Solvency II is expected to have on the daily life of pensioners over the next few decades via the new standards factored into the long-term guarantee measures used for the adjustment of the balance sheets of life insurers and pension schemes. In this regard, the authority of standards is all the more ambiguous in that they distance themselves from the market-consistent approach and are, at least, as much the result of horse-trading as of expert advice. As we saw in a number of cases, fixed rules and standards are impossible to define for frameworks of principles-based and risk-based regulation. Yet, the insurance industry fiercely struggled to ring-fence the most highly sophisticated standards and calibration formulas at the highest legal European

[22] Fair, safe, and stable insurance markets are at the core of the IAIS mission. See for instance: Address to the International Actuarial Association Council by Peter Braumüller, Chair of the IAIS Executive Committee, London, 13 September 2014. Available at: www.actuaries.org/LIBRARY/Presentations/2014/Braumuller_AddresstoIAACouncilLondonSaturdaySeptember13.pdf, accessed 15 July 2015.

[23] Interview with Doug Taylor, UK Financial Services Consumer Panel, London, 28 April 2015.

level. This was particularly true during the drafting of the Omnibus II Directive that was originally supposed to be not much more than an administrative exercise to bring the institutional environment of insurance regulation in line with the creation of EIOPA along with the new post-crisis European system of financial supervision. Finally, we should also take due account that the US regulator put additional emphasis on professional skills of insurance agents and brokers and accredited certification programmes to ensure a sufficient level of consumer protection. This distinct business education for qualifying the quality and security of insurance services sold by the industry clearly rests on societal values that support a less tangible influence of the United States on insurance markets (Kobrak, 2012). However, those skills are primarily assessed against an ability to comply with sophisticated standards which more often than not impose a market discipline comparable to invariable physical measures.

Third, regarding the space in which the authority of those regulatory standards is recognised, things have clearly changed over the last decade or so. Intra-state level regulation in the Unites States is losing ground with stronger oversight at the federal level, increasing convergence with Solvency II, and greater compliance to international standards such as those set by the IAIS. Likewise, Solvency II not only represents a sweeping change in setting standards for strengthening the unification of the insurance market under a common regulatory umbrella of the European Union. It also creates a global benchmark, often referred to as the new gold standard of insurance regulation. For their part, IAIS officials cannot state strongly enough that their work in progress is no replica of Solvency II and sets the stage for a truly global 'lingua franca' that should be 'clear, coherent, comparable and measurable' (Lezon, 2015: 55). Yet, according to Catherine Lezon, Vice General Secretary of IAIS for standards, 'IAIS's principles not only need adjustments with regard to Solvency II, but also, as the system is still very fragmented beyond Europe, to become applicable to other countries, such as Australia, Japan, Canada and the United States; norms will have to be transposed into domestic law with adjustments of various scales depending on each situation'.[24] A proper appraisal of the balance between the transnational projection of the first-mover advantage of Solvency II and the ability of other regulators within IAIS to make their voices heard is difficult to assess. It would indeed require an ex-post and item-by-item assessment after full implementation of Solvency II and the completion of the IAIS project in a

[24] Telephone interview with Catherine Lezon, Vice General Secretary of IAIS for standards, 14 July 2015.

number of years. For the time being and within the limits of this study, it is worth emphasising the ambiguous underpinning that remains at the core of the insurance capital standard (ICS) proposed by IAIS as part of its common framework for internationally active insurance groups. Although IAIS sets as its ultimate goal a common methodology by which the standard would achieve substantially the same outcome across jurisdiction, this is far from sure. According to the global consultancy firm KPMG, the ultimate form of the standards even remains 'worryingly unclear' (KPMG, 2015: 6). One of the main reasons for that is that US regulators were successful in their demand to keep generally accepted accounting principles (GAAP) used in the United States as an alternative valuation methodology alongside the so-called market-adjusted valuation approach developed by the IAIS for a globally comparable and risk-sensitive capital requirement standard. As a result, the IAIS is field-testing both options.[25] Another is the nature of safeguards set for the use of internal models from the insurance companies themselves for calculating the capital requirements – and therefore the extent of a system of compliance with non-state rules that encroach upon conflicting sources of transnational authority. The hybrid authority of insurance capital standards is thus caught between the principle of exclusiveness of territorial sovereignty in the United States and its continuing use of GAAP accounting standards and the inclusiveness of rules governing the global economy ultimately sought by the IAIS following the pioneer work of Solvency II. All in all, the spatial relations on which the effectiveness of regulatory standards for the insurance industry rests interpenetrate multiple jurisdictions. It remains highly ambiguous as to whether it is based on an endogenous logic of territorial sovereignty or the exogenous logic of the transnational underpinning of capitalism.

[25] For further details, see: International Association of Insurance Supervisors (2014, 2015).

6 Standards to Create New Insurance Markets

The preceding chapter shed light on the (re)insurance industry as an obscure, yet significant, object of global finance and governance, with a distinct focus on regulatory standards. This chapter goes a step further in looking at insurance standards used in market creation rather than associated to market regulation. How does the industry rely on standards to create new insurance markets? How do such standards help to transform the uncertainty of the material world into a fungible risk likely to be sold to prospective policy holders? Risk assessments and the drafting of sample insurance policies notably rest on complex procedures that seek to collect data that is as detailed and reliable as possible. To this end, insurers depend on the accessibility of such data, their comprehensiveness (their granularity in the jargon of the professionals), and last but not least, an industry-wide defined and harmonised format that can be easily exchanged and reported among all market players and regulators. As for any other data used in a service sector based on information, such interoperability requirements are reinforced by extensive use of ICT resources, longer value chains relying on all sorts of outsourced services, intra-firm exchanges between parent companies and their affiliates, as well as arm's length transactions on an increasingly global plane.

There is a consensus among our sources that the industry is known to use only a limited number of insurance-specific standards. Some instruments do, however, exist. The following stocktaking exercise aims to unveil a number of little-known standards that are nevertheless indispensable to the functioning of insurance markets. Some help to create new markets; others reinforce existing markets or drive changes in their functioning. To understand these two dimensions (market creation and market support or transformation), I shall examine two distinct areas. On the one hand, my enquiry focuses on how standards are instrumental in pushing the frontier of highly innovative and securitised insurance markets further, with a distinct focus on life insurance and its close connection to pensions. I thus first put the life insurance industry in the context of the challenges of the post-crisis environment, its relations with pension policy reforms, and on-going plans to strengthen the market integration of pensions. Against this backdrop, I present the project that insurers, pension schemes, and investment banks developed over several years for a standardised solution to pass over to capital markets the risk associated with longer and different expectations in populations' longevity – known as 'longevity risk'. On the other hand, I examine how existing markets heavily rely on standardised formats of data exchange. Here, the focus is on insurance against natural catastrophes, in particular the role of reinsurers that, together with states, are the only ones in a position to assume responsibility for covering rare but extreme losses resulting from natural catastrophes. After some background on generic data exchange formats widely used by, but not confined to, the insurance industry, I shall turn to the unique history of a standard developed over several decades by the world largest reinsurers to gain a more accurate picture of the exposure to natural hazards risks included in their portfolio. Finally, I shall look into another type of data exchange not confined to insurance: the standardised guidelines used for extra-financial reporting by the largest listed insurance and reinsurance companies around the world – the guidelines of the Global Reporting Initiative (GRI). While highly formalised with a view to establishing a harmonised – if not fully measurable – global standard of comparison for investors on globally integrated financial markets, such guidelines are closer to the societal pole of the material continuum of my topology of international standardisation. At first sight, this case may appear a long way off the core focus of my enquiry. It provides, however, evidence that standards are not sector-dependent and can set quality and security attributes even when they further oppose political economy objectives; it thus keeps on probing the extensive hypothesis put forward in Chapter 3.

The enquiry follows my basic three questions: who standardises what and where. In doing so, the evidence gathered will be wrapped up according to my three-dimensional framework. It will situate the actors setting those standards primarily along the private sphere of the institutional continuum. While what is standardised predominantly belongs to the physical pole of the material continuum, we will see that it nonetheless brings a number of societal issues on board. Finally, evidence gathered in this chapter suggests that although those standards largely belong to a logic of market creation and rationalisation, compliance remains ambiguous and falls short of a mere exogenous principle supporting the transnationalisation of capital accumulation.

The Cost of Not Dying

In the previous chapter we saw the crucial role played by standards in the paradigmatic shift of risk-based regulation as state regulators transfer a significant part of their authority to large insurance companies able to develop internal risk and solvency models, as well as to successfully lower the reserve capital and other requirements supposed to protect policy holders. Those models are assessed by regulators who ensure that the firms they supervise do not threaten the stability of the financial system and provide convincing reporting that an appropriate degree of protection has been undertaken for policyholders. For that, regulators and insurers need to agree on how to be sure that reserves match liabilities far into the future. That is obviously easier said than done, all the more so when insurers themselves lack an agreed methodology and calibration for standardised asset and liabilities pricing. This is what we now turn to, with particular focus on current developments in the life insurance market. The analysis sheds light on the technical specifications underlying the internationalisation of the life insurance market. In doing so, it keeps sight of their social and political implications following the global economic crisis, in particular with their close connection with financial services, contemporary pension policy reforms, and contentious plans to create a single market for pension funds within the EU.

Life Insurance after the Crisis

While securitisation was undoubtedly one of the drivers of the financial crisis of 2007–2008, the life insurance industry continues to assign it centre stage in the post-crisis environment. Shaping new standards for pricing securitised life insurance products and establishing commonly accepted contracts is critical in this regard. A standardised securitisation of life insurance products responds to three challenges of the post-crisis environment. First, it provides instruments of risk-based regulation that respond to attempts by state regulators to adopt a more complex and stringent regulation with closer convergence towards the banking industry – something we considered in the previous chapter focused on the insurance supervisory and regulatory environment. Then, it offers a convenient way to mitigate the dramatic implications that post-crisis ultra-low interest rates have for life insurance companies, facing a higher cost of their products (to match the loss of compound interests), lower returns from investments of their assets, and an increased valuation of their liabilities.[1] Set against the backdrop of a long-term prospect of super low interest rates, the pessimistic tone of the leading world reinsurer Swiss Re is largely shared among the industry: 'the longer interest rates stay low, the higher the losses in [life insurance] will be' (Swiss Re, 2012a: 38). Last but not least, the ability to scale up the market of securitised products according to standardised methods responds to the significant challenge that the long-term and macro trend of ageing has become for life insurance companies. The impact of an ageing population varies according to the type of pension arrangements. The evolution of fertility rates, improved life expectancy, and the end of the baby boom generation have joined market ideology as main driving forces behind the shift towards a massive substitution of defined contribution for defined benefits systems across industrialised countries. Significant tax and other state incentives also support the development of complementary funded private pensions. Since the crisis of the 1970s, debates on the so-called burden of social expenditures and more broadly the crisis of the welfare state have spread across countries through various transnational channels to 'become staple items on the political agenda' (Leimgruber, 2013: 293).[2] Governments have repeatedly attempted to push through vast reforms to close the funding gap between contributions and benefits.

In the post-crisis environment of low interest rates and risk-based regulation, the life insurance industry can surely play its own game in the reforms of pension systems swiping countries with an ageing

[1] In an environment marked by a long-term prospect of low interest rates, the price of life insurance premiums goes up as a lower share of the benefit sold by the policy is expected to be funded by compound interest rates. For instance, at 0 per cent interest rate, a benefit of $100,000 in twenty years would require payment of a yearly $5,000 premium, whereas with a 5 per cent investment return this would only require an annual payment of $2,880, with 42 per cent of the benefit paid out of interest rate income. Low interest rates thus make life insurance products either more expensive or their benefits lower, and this clearly affects the demand for insurance policies. As insurers invest most of their premiums in high-quality bonds, low interest rates also reduce their investment returns. Finally, lower interest rates increase the value of their liabilities. Following the previous example of an insurer with a liability to pay someone $100,000 in twenty years' time, the value of that liability today must be discounted by the expected amount derived from compound interests over those next twenty years. The present value of the future amount is thus reduced in proportion to the average interest rate expected for that duration. The smaller the interest rate, the higher the value of the future sum in today's money – that is, the higher the liability weighs on their balance sheet. For further details, see: Swiss Re (2012a).

[2] For insights on the role of international organisations such as the OECD and the World Bank and other transnational policy actors on the privatisation of pension policies and the shift towards transferring risks to policy holders, see, among others: Orenstein (2008) and Mandin and Palier (2009).

population. Not only does it guarantee against the cost of dying (i.e. paying an indemnity to a beneficiary in case of death of the insured), it also sells policies to hedge the cost of not dying (i.e. providing pay-outs to the insured for an agreed period of time, sometimes as long as the time s/he stays alive). The cost of not dying hedged by life insurers is thus closely related to the annuity market (Inkmann et al., 2011: 281). Annuities are generally defined as contracts that provide periodic payments for an agreed-upon span of time. With substantial variations in length-of-life across populations, a life annuity allows a retiree to exchange either an accumulated capital or a lump-sum for a guaranteed stream of income that will be paid as long as she is alive (Brown et al., 2001). The development of standardised instruments for creating a new global market of securitised pension-related policies thus rests on a proper understanding of the risk borne by not dying, how to price it, and of course, in which market to expect most revenues.

While the United States remains by far the biggest country in terms of pension funds' assets under management (with close to 60 per cent of the estimated $25trillion in OECD countries), the United Kingdom is by far the largest market for annuities. This is so since the accumulated capital of occupational plans and personal pensions must be used to purchase an annuity at retirement. Until the conservative Chancellor George Osborne ended compulsory annuitisation in 2014 – a reform labelled as the biggest of the century by asset managers at JP Morgan (Berens, 2015) – life insurance companies operating in the UK not only benefited from the world's largest market but led in product innovation and ways of developing risk differentiation (Rusconi, 2008; Marschallek, 2011).

Longevity Risk and the Design of Lifemetrics

In the profession, the risk hedged by financial instruments that pass the securitised solutions imagined by insurers to offload their ageing and pension-related risk over to capital markets is known as longevity risk. The notion was forged around the turn of the century to deal with the birth of those risk transfer markets. Longevity risk is thus related to the 'uncertainty surrounding the increases in life expectancy— as a result of unanticipated changes in mortality rates' (Blake et al., 2013: 5). Accordingly, it does not seek to address the viability of pension systems or solvency of insurers per se, but rather the complicated issues that arise when insurers, pension funds, pension schemes, and investment bankers seek to hedge the risk associated with the fact of guaranteeing continued streams of revenue to different populations that will experience different longevity outcomes. For all those actors involved in this new 'life market'

(Blake et al., 2013), the cost of not dying is so difficult to price that it needs standards against which to define the market. According to one leading expert who helped to give major currency to the notion, longevity risk is 'the most important risk that pension funds and insurers face, because it is the only one you can't hedge – in contrast to credit or interest risks using well-known financial models – and it is the most unfair towards future generation that would take the burden of it if not properly addressed now'.[3]

Over the last decade, insurance services were part and parcel of the surge of buy-out arrangements, annuity contracts, and securitised solutions sold to pension funds and pension schemes to offload the longevity risk borne on their balance sheet.[4] It remains difficult to have reliable estimates in the five leading markets (UK, United States, Netherlands, Canada, and Ireland) due to a lack of transparency and comparability in the information released by large consulting firms advising and tracking those deals. The last few years typically saw some jumbo deals of over £1 billion in each country, with many smaller deals. Figure 6.1 presents an overall picture of the growing market of longevity risk transfers in the United Kingdom since the outburst of the global financial crisis. The peak of 2014 can be largely explained by the large deals done prior to Solvency II coming into force.

Despite such recent developments, life insurance and reinsurance companies have experienced difficulties in creating bold new markets in relation to an ageing population and current reforms of pension policies. The lack of standards for pricing the cost of not dying was from the outset the main difficulty faced by the industry. Why? A first response is to consider that what is true for financialised capitalism is also true for the securitisation of insurance. Without uniform contract and pricing standards, capital markets cannot expect to attain the depth and liquidity

[3] Interview with David Blake, Director of the Pension Institute, Cass Business School, London, 20 April 2015.

[4] In a pension buy-out, a pension fund and/or plan sponsor hands over all the assets and liabilities of the fund to an external provider, typically an insurer or reinsurer, who then has the sole responsibility for making payments to the members of the pension plan or fund. As emphasised by an OECD report, 'while the plan sponsor offloads all risk, this arrangement exposes plan members to counterparty risk, or the risk that the insurer becomes insolvent, as the structure no longer has the same benefit protection mechanisms in place as the pension plan' (OECD, 2014a: 177). The situation is different with a pension buy-in, in which the pension fund or plan sponsor buys an annuity contract to rely on (re)insurers to fully or partially insure its liabilities, while retaining them and remaining responsible for the payment of pension benefits to its members. In both cases, the use of capital market to furthermore hedge those contracts has dramatically surged in the aftermath of the financial crisis and the prospect of long-term, super low interest rates.

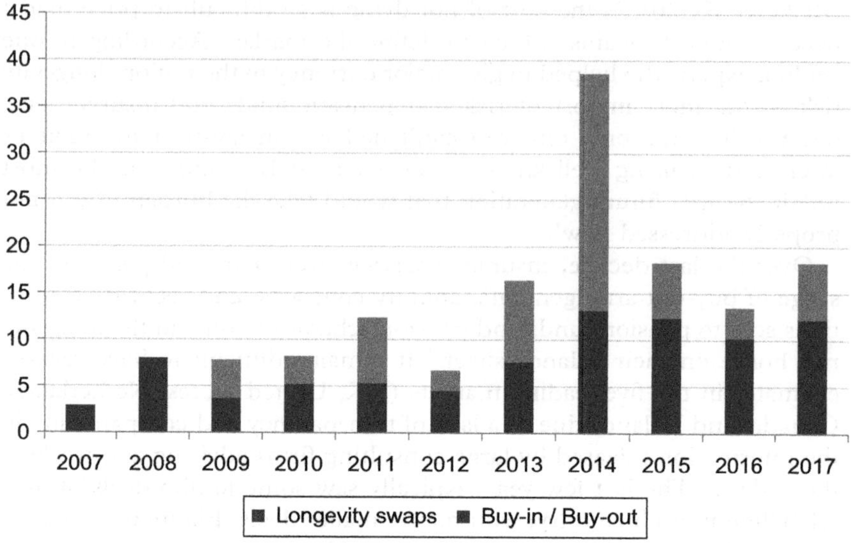

Figure 6.1 Volume of longevity risk transfer deals in the United Kingdom (2007–2017; £bn).
Source Hymans Robertson (2018).

required to scale up from a niche financial innovation (Lysandrou, 2016). Standardised forms of provision are requested whenever a financial market grows in scale; they assist asset managers' demands for systematic comparisons of securities in determining their suitability for inclusion in a particular portfolio. While life insurers have over centuries developed sophisticated products using mortality tables, the securitisation of those products generates additional requirements in terms of standardised bases of reference. A second answer – more specific to the insurance industry – is thus required to reconstruct the origins and developments of standards supporting the securitisation of longevity risk and so-called life markets.

In the early 2000s, the idea of developing a standardised longevity risk index had been in the air for a few years. Longevity capital markets were seen as potentially relevant for the banking industry working more and more closely with pension funds in order to develop packaged investments and hedging instruments. Swiss Re (then, the largest reinsurer of the world) inaugurated the first generation of capital markets instruments in December 2003 with the issuance of a so-called mortality bond known as Vita 1 (i.e. the name of the special purpose vehicle created for that). But the instrument merely transferred the model previously used for natural catastrophe bonds: it only reduced exposure to catastrophic mortality events such as a severe outbreak of influenza, a major terrorist attack using weapons of mass destruction, or a natural catastrophe (Blake et al., 2013: 15–16).[5] Together with experts from Heriot-Watt University in Edinburgh, the Cass Business School Pension Institute founded by David Blake had bigger plans for scaling up the market. In 2005 it organised the First International Conference on Longevity Risk and Capital Market Solutions, which would hereafter take place annually. Together with colleagues, the objective was to ensure not only the hugely complicated maths of the new market but also to understand how to design standardised contracts that would respond to the difficulties identified in the first issuance of bonds.

The creation of new capital market instruments cannot expect long-term viability without meeting the needs of both the hedgers (those buying financial instruments that cover the risk; e.g. an insurer, a pension fund, or a pension scheme with too high a liability related to current or future annuities) and the speculators (those selling the instrument; e.g. an investment bank, usually with the support of a large insurance consultant firm). Whilst the former look for hedge effectiveness, the latter seek liquidity like any other financial actor. Yet, a liquid market in which hedging instruments can be easily exchanged depends on standardised contracts whose form and substance are intelligible and comparable to all actual and potential market actors. As Blake and colleagues emphasise, 'the fewer the number of standardized contracts traded, the greater the potential liquidity in each contract, but the lower the potential hedge effectiveness. There is therefore an important trade-off to be made, such that the number of standardized contracts traded provides both adequate hedge effectiveness and adequate liquidity' (Blake et al., 2013: 12). The standardisation of longevity risk indices is thus caught in that tension between standardised index-based hedges and customised hedges. Standardised contracts have the advantages of simplicity, cost, and liquidity. In their simplest form, they support an index-based longevity swap (a derivative) involving a payment to the pension scheme or insurer based on the longevity experience of a reference index. Yet, understanding 'how good' the risk reduction is remains a difficult problem as the referred index will never exactly match the actual annuity payments being made by the insurer or pension scheme (Cass Business School and Hymans Robertson LLP, 2014: 8). Guy Coughlan, then newly

[5] For an analysis of this longevity bond market from a poststructuralist approach focused on the particular understandings of time that it enshrines to produce truth-base insurable events, see: Lobo-Guerrero (2014: 54–71).

appointed head of the asset liability management (ALM) risk team of JP Morgan in London, was also present at the creation and shared the understanding that 'an essential requirement for creating any new liquid market is standardization'.[6] The creation of a liquid market would thus require 'a standardized index ... as an unbiased reference by all participants [and] a limited number of standardized contracts in which liquidity can be concentrated' (Coughlan et al., 2007: 4). In his view, in its early stage, the market could be built around just eight standardised contracts with a specific maturity (e.g. ten years), two genders (male, female), and four age groups (50–59, 60–69, 70–79, 80–89).

It is within this mind-set that the Lifemetrics initiative began at JP Morgan London in early 2007 to provide an effective long-term hedge of the longevity risk of a pension plan or annuity portfolio.[7] The rationale from the start was that standardisation was necessary to reach scale, support liquidity, and expect growth of the market with proper intermediation between buyers and sellers. Coughlan approached Swiss Re to set up a joint association bringing the major players among insurers, banks, pension funds, and investors together. In April 2011 JP Morgan thought that a critical mass was reached and deemed it worthy of transferring the Lifemetrics initiative and related longevity standards to the Life and Longevity Markets Association (LLMA), a not-for-profit venture established for that purpose. Interestingly, the establishment of suitable and consistent standards, conventions, and best practices are an integral part of its objectives in the promotion of a liquid traded market in longevity and mortality-related risk. In August 2018, LLMA membership included Aviva, Axa, Deutsche Bank, JP Morgan, Morgan Stanley, Prudential Plc, and Swiss Re, to whom it provides historic and current indexes of mortality rates and period life expectancy levels across various ages for the four largest markets that are the United States, England and Wales, the Netherlands, and Germany. It furthermore provides standardised valuation models for longevity and templates for standardised derivatives such as so-called q and s forwards. According to experts close to the field, Lifemetrics standards developed by LLMA are

6 Interview with Guy Coughlan, Chief Financial Risk Officer, USS Ltd, and former head of the asset liability management (ALM) risk team of JP Morgan, London, 30 April 2015.

7 Interview with Guy Coughlan, Chief Financial Risk Officer, USS Ltd, and former head of the asset liability management (ALM) risk team of JP Morgan, London, 30 April 2015; interview with Pretty Sagoo, Director, European Insurance Risk and Capital Solutions, Deutsche Bank, and Director Board LLMA & Chair LLMA and IFoA Joint Longevity Basis Risk Working Group, London, 28 April 2015.

considered to have no competitors on the market even if new refined methodologies are developed by practitioners elsewhere.[8]

Although slow to take off and having not yet gathered pace to reach the full cruising speed of mature markets, standards supporting the issuance of securities on longevity risk have nevertheless accomplished a long journey since their early days of discussion in the academic circles of Heriot-Watt University in Edinburgh and the Cass Business School Pension Institute in London, as well as among large insurance and pension consultants in the United Kingdom, such as Aon Hewitt, Mercer, and Hymans Robertson. It is particularly worth noting that it has now gained a highly coveted prominence in OECD publications. In 2014 the OECD Working Party on Private Pensions – well known for its role in promoting the three pillar system – released a comprehensive report on longevity risk. The report emphasises in particular that 'Index-based instruments offer a solution to the constraints of capital markets investors in supplying longevity protection ... further development of these instruments could be facilitated by additional standardization and transparency in the market' (OECD, 2014a: 183). What is more, the 2014 issue of the OECD flagship publication on pensions put longevity risk in its first chapter. In this finely tuned analysis of far-reaching challenges of pension systems in the low returns, low interest rates, and low growth environment of the post-crisis era, standardisation is portrayed as a key tool of longevity risk management: 'Capital markets may have the potential to provide additional capacity if standardised instruments to hedge longevity risk via longevity bonds, swaps and other derivative contracts were available. For purposes of standardisation, these instruments may need to use longevity indices based on the general population' (OECD, 2014b: 39).

In the previous chapter, we saw that the private insurance industry expresses interest in a regulation-light approach and opportunities for expanding access to a European-wide market in the making. Lifemetrics is undoubtedly a market standard far away from any form of regulatory standards. It is situated on the private, technical, and transnational poles of my standardisation topology. Yet, the authority of the standard remains ambiguous. States are not necessarily excluded from the creation of such a new market. The potential role of governments in supporting the standard

[8] So far, the only competitor on the market is the Xpect - Club Vita Indice, a more detailed series of longevity indices tailored for England and Wales by Club Vita, Deutsche Börse, and Hymans Robertson's longevity analytics arm. Cf. aforementioned interviews; www.llma.org, accessed on 21 April 2015; 'Deutsche Börse and Club Vita to launch new indices for pension schemes pursuing index-based longevity swaps', Deutsche Börse Press Release, 15 March 2012.

remains, indeed, a disputed issue. According to Blake, governments have an important role to play and should take an active part in it: only they have access to the information needed to help with the construction of sophisticated national longevity indices; moreover, as longevity risk is not actively traded in the capital markets, governments are trusted as important enablers of capital market development if they issued themselves longevity bonds that would facilitate price discovery (Blake et al., 2014: 264). In the same vein, Swiss Re (which was associated with the project right from the beginning) views the viability of the instrument as possible only through massive state involvement that would help define a still-lacking reference price. State issuance of such bonds, the argument goes, would 'encourage the development of the market … and facilitate private companies offering similar products' (Swiss Re, 2011: 6). Others, on the contrary, share a more fundamentalist view of the market and do not see why governments would have any role, especially when they have their own longevity risks to solve in the first place, with massive defined benefits pensions schemes harder than ever to fund, let alone quantify their liabilities.[9] Moving to the second dimension of my analytical framework, we can appreciate how standards setting longevity indices stand at the technical end of the material continuum, even without entering the sophisticated maths of Lifemetrics. However, this is not unambiguous in terms of conveyed social values. All sorts of assumptions are made on how detailed a differentiation can be set among groups of population. The whole exercise is also posited on the political economy assumption that liquid capital markets instruments are the best guarantor of long-term revenues to an ageing population. Finally, regarding the spatial spectrum of standards' recognition, the longevity standard was developed against the backdrop of the specificity of the annuity market for life insurance companies in the United Kingdom. From the outset, it was developed as an instrument ready for tapping the other major annuities market around the world, in particular those of the United States and the Netherlands, together with the rising German market resulting from the early 2000s so-called Riester reforms. More generally, an important lesson to draw is that the standardisation of an atypical service industry such as life insurance is paved with difficulties, despite the fact that it does not face the usual challenges resulting from highly relational and immaterial activities generally seen in the literature as enabling standardisation and internationalisation rather than the contrary. From this point of view, there is no sector-specific

[9] Interview with Pretty Sagoo, Director, European Insurance Risk and Capital Solutions, Deutsche Bank, and Director of the Board of LLMA & Chair of LLMA and IFoA Joint Longevity Basis Risk Working Group, London, 28 April 2015.

explanation in the lack or prospects of standardisation likely to support or hinder the expansion of the tertiary sector on an international, let alone global, scale. First of all, standards support distinct, and sometimes opposing, conceptions of the market. In the life market for longevity risk, standardised contracts and pricing support a securitisation of the insurance and pension industry with the development of derivatives on the capital market, in contrast to more customised and hedging techniques defined on a national basis and used for centuries by actuaries hired by insurance companies.

At the Heart of (Re)Insurance Standards

On 11 March 2011, an earthquake with a magnitude of 9.0 struck Japan and triggered a powerful tsunami that caused the death of around 20,000 people and widespread damage to infrastructure and property, including the nuclear power station of Fukushima Daiichi, with a meltdown of three of its six reactors. According to industry experts, this was the most costly natural catastrophe of all times, with the highest insured losses ever recorded for an earthquake (Swiss Re, 2012b). Although reinsurers were still cautious with estimates almost a year after the catastrophe, a remarkable thing behind the headlines is that, within days, the world's largest catastrophe risk modelling companies were able to put forward detailed and reliable figures of incurred losses.[10] Estimates did vary in a proportion of one to three, with the highest figures reaching US$300 billion or around 5 per cent of the GDP of what still was the world's second largest economy. Since then, those figures have not changed dramatically, with economic losses estimated between US$210 and US$300 billion. In view of the far-induced paralysis in which the country was at the time, how was it ever possible to provide numbers so quickly and precisely? Moreover, with the ability to provide such reliable figures so quickly, how can we explain a relatively high proportion of one to three in their variation? As we will see in this section, natural catastrophe risk exposure data exchange standards played a prominent role in this regard. Similarly, divergent methodologies and assumptions (such as exclusion clauses of nuclear risks) used by risk modellers go a long way towards explaining some discrepancy in the figures. The section begins with some background on generic data exchange formats. The unprecedented history of a global standard for natural catastrophe risk exposure set by large reinsurers will follow. After that, my focus will widen again to data

[10] 'Cost to Insurers forecast to rival hurricane Katrina', *Financial Times*, 20 March 2011.

exchange standards not limited to a defined branch of the insurance industry with a study of the extra-financial reporting guidelines used by the largest (re)insurance companies of the world. The evidence gathered suggests that although standards supporting insurance market creation and intermediation predominantly rest on the private, physical, and transnational segments of our typology, some of them can nevertheless include a slightly more societal dimension, such as with the case of the comprehensive sustainability information reported with the use of the Global Reporting Initiative (GRI) guidelines.

Exchange Data

The insurance and reinsurance industry relies on data exchange formats used in many other information-based service activities (e.g. in the domain of finance as well as auction-driven markets and IT services). A set of internationally agreed standards, directories, and guidelines for the electronic interchange of structured data has been defined to facilitate business practices between independent, computerised information systems. Most of these standards are based on the universal Extensible Markup Language, better known as XML. This set of rules for encoding documents in a computerised form was developed in the late 1990s and early 2000s by the World Wide Web Consortium (W3C), the institutional platform which includes more than three hundred firms, computing departments from universities, and publicly funded research centres, ministries, and community representatives working together for promoting open source and open standards for the Web. The development of this language in the 1990s marked a shift in computer science as its extendibility made it possible to store and share any kind of data. Many office suites software rely on it.[11] While the W3C epitomises the significance of transnational

[11] I analysed in detail elsewhere with colleagues how the XML provides an outstanding case study of commodification of service standards with broad implications for the global computer services market. The study highlights how the largest multinational corporations pay special attention to gain a recognised international standard for such a major technological innovation. It shows how the XML standardisation processes affected market structure and led to market capture, in particular through the strategic use that Microsoft made of negotiation arenas. While the ISO had already adopted an open source standard set by IBM and Sun Microsystems, Microsoft was successful in making its own technical solution a recognised ISO standard as well (ISO/IEC 29500–1:2008. Information technology – Document description and processing languages – Office Open XML File Formats – Part 1: Fundamentals and Markup Language Reference). A broader lesson to be drawn from the case is that XML standardisation also helped to establish a distinct model of information technology services at the very expense of the monopoly on proprietary software defended by Microsoft and successfully sanctioned by the ISO. For further detail, see: Vion et al. (2013).

private technical governance platforms including a relatively broad array of civil society stakeholders on diverse issues pertaining to ICT and the use of the Internet, the use of standardised formats of data exchange in numerous industries is also supported by a body that is a priori more strictly public. It is indeed within the framework of the Economic Commission for Europe (ECE) of the United Nations that the Centre for Trade Facilitation and Electronic Business (UN/CEFACT) developed a first set of interchange rules in the form of 'Guidelines for Trade Data Interchange' (GTDI) that were subsequently published in 1981. The next stage in the work towards a common universal set of interchange rules for trade data was the development of the United Nations Electronic Data Interchange for Administration, Commerce and Transport (UN/EDIFACT) syntax rules.[12] This syntax has not only been taken over in the international standard ISO 9735, but has also become an integral part of the United Nations Trade Data Interchange Directory (UNTDID), which establishes how messages must be structured on a set of functional modules. Similar procedures are used in the banking industry and are better known to the public. Anyone who has wired some money abroad has at least heard of BIC codes, i.e. the business identifier code used by financial and non-financial institutions to facilitate automated processing of information for financial services. If not, they may instead have heard of SWIFT codes, those same codes handled by the Society for Worldwide Interbank Financial Telecommunication (SWIFT) based in Belgium. Few of us know, however, that those tools used for addressing messages, routing business transactions, and identifying business parties are all part of the international standard ISO 9362.

Although formally located within a UN body, the input for those technical specifications is for the most part driven by the private sector. Insurance data exchange formats elaborated within the UN/CEFACT rely heavily upon the expertise provided by the eEG7, the European forum for the development of e-business standards for electronic communication in the insurance sector. More recently, ACORD (Association for Cooperative Operations Research and Development) has provided further input to the UN/CEFACT agenda.[13] In contrast to

[12] United Nations Centre for Trade Facilitation and Electronic Business, UN/EDIFACT Draft Directory, www.unece.org/trade/untdid/texts/d100_d.htm, accessed 31 August 2010; for further detail, see the following UN/EDICAFT main webpage: www.unece.org/cefact/edifact/welcome.html accessed 30 July 2015.

[13] ACORD regularly organises joint events with UN/CEFACT. I took part as an observer to the one organised on 31 August 2010, called Insurance Vision Day, during the 17th Forum of the United Nations Centre for Trade Facilitation and Electronic Business (UN/CEFACT).

eEG7, ACORD's membership is closer to the American and, increasingly, the London market. It therefore pushes for a more global reach in the elaboration of specifications likely to facilitate the development of data and format standards in the insurance industry. One of its major achievements so far is the development of an Insurance Core Component Library as a subset of the global UN/CEFACT Core Component Library. The standards published target various business lines, such as individual and commercial property claims, and commercial. They provide the necessary requirements for a proper transfer of information between policyholders, professional intermediaries, insurers, and other involved parties; they support the establishment and management of insurance contracts, the handling of claims, and accounting practices.[14] Data exchange formats undertaken by eEG7 and ACORD under the UN/CEFACT are clearly positioned in the most private, technical, and exogenous subdivision of the framework of conceivable international standards. Since its creation in 1970, ACORD has successfully positioned itself as a prominent actor of generic data exchange format standards for the insurance industry, first in the United States and now increasingly on the global scale. We will soon see that it has recently become an important actor as well of data exchange standards in the distinct field of reinsurance for natural hazards.

Reinsuring NatCat

Putting aside generic data exchange formats, the few standards specifically dedicated to the insurance industry remain strictly private and oligopolistic, and defined on a narrowly technical basis. Standards developed to make a realistic assessment of risks relating to natural hazards are a good case in point. Insurance losses caused by natural catastrophes have risen dramatically over the last thirty years. According to extensive data collected by Swiss Re, the ten-year average total economic losses are a multiple of five of what they were in 1990. In addition to the impact of global warming, economic development, and population growth, a higher concentration of assets in exposed areas keep increasing the economic cost of natural disasters (Swiss Re, 2015: 6). In 2017 alone, natural catastrophe-related economic losses were around $330 billion, with claims close to $138 billion, significantly above the previous ten-year annual average ($50 billion). With the quick succession of hurricanes Harvey, Irma, and Maria counting for two-thirds of insured

[14] Interview with Jürgen Heck, Program Director for Europe, ACORD, Zurich, 4 June 2010. For further detail, see the following website: www.eeg7.org.

losses worldwide, this was the highest level ever recorded in a single year (Swiss Re, 2018a: 3–5). The insurance and reinsurance industry is therefore well advised to use agreed methods in collecting the information on which to base the evaluation of the financial fallout that such events can have on their portfolio.

The uncertainty surrounding the occurrence of natural catastrophes, in particular earthquakes, their infrequency, and the great fluctuation of events, whose consequences can be devastating but not necessarily insured, are among the factors that make insurance in the field of natural catastrophes extremely complex. Indeed, how could it ever be possible to evaluate risks and hedge them with a reliable level of precision under such circumstances? Unpredictability, high amplitude, and geographical concentration may indeed make the calculation difficult, but not impossible. That is precisely the job of insurers! They usually build their natural hazard models by classifying four different sets of data (Swiss Re, 2003: 11–37). First, information regarding the *hazards* themselves, i.e. where, how often, and with what intensity do events occur? Second, insurers collect material on *vulnerability*, i.e. what is the extent of damage at a given event intensity? Third, they need data on the *value distribution*, which will determinate the geographical localisation, the nature, and the value of insured objects. Last but not least, detailed *insurance conditions* included in policies fix the proportion at which the loss is insured. Those four factors are combined in the process of estimating potential losses resulting from natural catastrophes, whose cost is assumed in varied proportions by the policy holders through the premium paid, the insurer guaranteeing the risk, the reinsurer to whom this risk is (partially) ceded, and, in certain cases, even the state, which can act as a security provider of last resort for a risk too high to be covered by reinsurers (as is the case in Japan or New Zealand).

Natural catastrophe risks confer a particular role to reinsurers. Apart from states, they are the only ones solid enough to hedge the risks that result from hazards as rare as they are extreme. In contrast, insurers are usually unable to hedge enough capital and their portfolio is insufficiently diversified to cover such risks, which accordingly are in part or totally ceded to reinsurers. For his part, the reinsurer goes beyond a simple evaluation in matching risk with the highest possible accuracy, diversifying his portfolio, and using additional securitised products available in global financial markets. He must also control the accumulation of risks included in his portfolio of reinsured policies and thus avoid too big a concentration on one type of hazard, geographical localisation, vulnerability, or insured objects. The recording of data on loss-exposed values in the reinsurance of natural hazards is known as 'accumulation control'. According to

Swiss Re, the world's second largest reinsurer that, together with Munich Re, controls more than a third of the world market, accurate accumulation control is an essential precondition for arriving at a meaningful assessment of the financial risk involved in insuring natural hazards (Swiss Re, 2003: 13). The inventory of the whole policy portfolio of a reinsurer will be reliable only if it can count on accessible and relevant data that are reported in the most harmonised and interoperable way.

This is where the use of internationally recognised standards for data exchange of risk exposure to natural catastrophes becomes important. Yet, the insurance industry is well known for its low level of coordination and the limited number of standards set by its major players to better structure the market. For a long time, the only standard available at the international level for data collection and exchange in the domain of accumulation control of risks of natural hazards was based on the geographical classification of so-called Cresta zones.[15]

Cresta (Catastrophe Risk Evaluating and Standardizing Target Accumulation) dates back to the late 1970s, when the two giants of the reinsurance industry, Swiss Re and Munich Re, together with four smaller European reinsurance companies, organised several meetings to harmonise the zoning of natural catastrophe risks in order to evaluate the accumulated risk in insurance policy portfolios underwritten in bulk by reinsurers. Heavy losses related to large earthquakes in Managua (Nicaragua) in 1972 and in Guatemala in 1976 came to them as a surprise, as they had no idea whatsoever of the exact location of the risk

[15] The acronym CRESTA has several meanings. It originated in 1976 as the name of the hotel where the first meeting of reinsurers took place on this issue in the little Swiss resort of Savognin in the canton of Grisons. The gathering quickly found the following set of relevant words instead: *Cooperation of Reinsurers for EQ Studies and Tariff Analysis*. However, reference made to tariff analysis turned out to be tricky later on in regard to existing American antitrust laws. The current meaning was defined in the late 1990s and refers to *Catastrophe Risk Evaluating and Standardizing Target Accumulation*. Sources used for the present account of Cresta origins and its more recent evolution come from the following interviews to which (for the purpose of clarity) I will not systematically refer hereafter: Ernst Leffelaar and Thomas Grollmann, GenRe, Köln, Germany, 19 February 2010; Rudolf von Flüe, Swiss Re pensioner, Rüschlikon (Zurich), 11 June 2010; Peter Hausmann and Christina Schlenther, Swiss Re, Zurich, 4 June 2010; Jürgen Heck, ACORD, Zurich, 4 June 2010; Yörn Tatge, AIR, Rüschlikon (Zürich), 10 June 2010; Peter Beresford, EQECAT, Rüschlikon (Zurich), 11 June 2010; David Carttar and Jeff Kilbreth, RMS, Rüschlikon (Zurich), 11 June 2010. Further information results from my own (or my assistants') participatory observations at two consecutive Cresta general meetings with around forty experts under the aegis of the Cresta Secretariat assumed alternatively by Swiss Re and Munich Re (Cresta meetings of 22–23 June 2007 and 10–11 June 2010 at the Swiss Re Centre for Global Dialogue, Rüschlikon (Zurich)). Relevant websites and professional magazines provide further subsidiary sources (for instance: Gusman, 2010).

exposure included in their contracts. As emphasised by a participant to the first meeting that took place in 1976, 'The reason was to gather standardized earthquake data. And this was a common interest for the whole insurance and reinsurance industry, including local insurance bodies.'[16] Over the next twenty-five years Cresta became the most important zoning format for reporting natural catastrophe exposures in the insurance and reinsurance industry. While the standard was first focused on earthquakes and contracts pertaining to Latin American countries, it reached the European market in the early 1990s in the aftermath of disastrous floods across the continent. Soon afterwards, the increased losses from tropical storms in the United States and elsewhere drove an increasingly global reach of the private zoning standard set by Cresta. At the end of the 1990s, relevant data were collected in more than 70 countries; 326 insurance and reinsurance companies subscribed to the standard in Europe, 76 in the United States, 13 in Canada, 40 in Asia, and a few dozen elsewhere.[17]

Although mainly focused on the definition of harmonised geographical zoning – and thus, according to my analytical framework, situated at the physical end of the material continuum of standardisation – Cresta codes also used this spatial expansion to broaden the catalogue of collected data, some of which clearly included greater societal concerns. Thus, data collected would not simply be about the location, the number, and the value of the insured objects but also on the quality of the building material, the type of their occupancy, their content, and so forth. At the same time, risk exposure data exchange formats provided by Cresta expanded to natural hazard risks other than earthquakes, in particular those that predominantly concerned the new areas included in codes (such as floods and storms for Europe as early as the mid-1990s). In 2003, Swiss Re could claim with confidence that Cresta zones were 'widely recognised as the global standard for the geographical breakdown of insurance data throughout the insurance sector' (Swiss Re, 2003: 22). Yet, in spite of such successful developments, Cresta still remained largely a standard of geographical zoning that lacked detail and was poorly adapted to the largest market in the world, the policies that include the risk of windstorms along the Eastern and Southern coasts of the United States. With a level of aggregation still very high – approximately that of a French department – it was of little help for modelling on a reliable basis potential losses incurred from insurance policies included in portfolios in dozens of millions.

[16] Interview with Rudolf von Flüe, Swiss Re pensioner, 11 June 2010, Rüschlikon (Zürich).
[17] Interview with Ernst Leffelaar, Gen Re, Köln, 19 February 2010.

Since then, the evolution of Cresta has largely remained in the hands of its duopolistic founding fathers, Swiss Re and Munich Re. It has, however, faced increasing competition from new entrants. Beginning in the early 1990s, three highly innovative companies of risk modelling software and consulting services (RMS, AIR, and Eqecat) created a new niche market. Known as model providers, they developed catastrophe risk management models based upon innovative applications of mathematic, computing, and geographical methods. Even now, the world market of catastrophe model providers is mostly shared between the trio of RMS (leading the market), Eqecat, and AIR Worldwide. While data gathered by Cresta were based on paper forms up to 1998, those models were from the outset highly computerised and provided detailed and dynamic zoning information.

Without any doubt, model providers are the new players that have challenged a market previously in the hands of the two giants of the reinsurance industry. Each of them developed their own format. While the EDM format provided by RMS is registered as a patented technology (like the Eqecat format) with closely monitored licenced usage, it has become so common throughout the world that it is often considered a *de facto* standard. The Universal Cession Electronic Data Exchange (UNICEDE) format developed by AIR Worldwide is also proprietary, but in contrast to EDM it is freely available. Continuous progress in modelling technologies constantly seeks access to more data, ever more detailed and diverse. According to Jeff Kilbreth, Senior Vice President for Software Product Management at the market leader RMS, around 95–98 per cent of risk analyses are undertaken at a detailed level in the United States, with figures reaching 60–65 per cent in Europe and 20–25 per cent in developing countries. In his view, this looks like 'a worldwide journey towards getting better at mastering detailed data'.[18]

For fifteen years, Cresta competed fiercely with this much more effective method for building harmonised databases of risk exposure to natural catastrophes. The modellers provided cheaper, highly computerised, and more detailed information. As one director of the trio of model providers points out, 'with the cat modellers coming to the market, there was definitely the need to have more high resolution Cresta zones; and that's exactly what's happening'.[19] Less detailed data at a higher level of aggregation remain relevant for reinsurers' basic 'accumulation control' and

[18] Interview with David Carttar and Jeff Kilbreth, RMS, Rüschlikon (Zurich), 11 June 2010.

[19] Interview with Yörn Tatge, Managing Director, AIR Worldwide, Rüschlikon (Zürich), 10 June 2010.

for developing countries, where detailed data are available for not much more than 20 per cent of risks covered. Yet, access to interoperable and detailed data has become part and parcel of the analytical work carried out for industrialised and emerging countries by model providers, as their specialised services are used and aggregated by insurance brokers, as well as more complex tasks undertaken by reinsurers. The following account of one the pioneers of the first Cresta zoning is clear evidence of the shift that has taken place: "At the beginning of the Cresta standard, we wanted to have the sum of sums insured per zone; so it was aggregated on a very high level. Nowadays, and because many policies are very sophisticated … this is not enough; so now we want to get the information on a much more detailed basis; we drill down, we go much more to the original policy information'.[20] Here the standardisation of accumulation control of risks related to natural catastrophes encounters once again ACORD (Association for Cooperative Operations Research and Development) – the consortium of (re)insurers that supports the development of data and format standards in the insurance industry.

In the early 2000s, the three leading international model providers, RMS, Eqecat, and AIR, reached a size that placed them in a position to negate the previously undisputed market power of Swiss Re and Munich Re. The challenge is to develop new zoning standards with uniform and detailed risk exposure data on natural hazards for the whole industry. Against this backdrop, ACORD established a working group on catastrophe exposure data standards. After more than a decade of fierce competition among data formats and online input templates, a focused and collective action of standardisation has taken place. The first two versions of the standard were published in 2003 and 2006, albeit without much success. A new working group was established in 2009, this time with all prominent actors of the industry, in particular the world's major (re)insurers and the trio of model providers around the same table (Davis and Garda, 2009). But before that, it took a little while to persuade RMS to join in. As market leader, it thought it had no reason to join this collective endeavour. According to an expert who took part in this new working group, 'typically they were absent; we invited them a couple of times and it was tough to bring them on board'.[21] Eventually, the project to develop the standard in relation to the creation of an innovative IT platform supporting highly value-added consultancy services convinced them to join forces. The agenda of this new NatCat working group is

[20] Interview with Ernst Leffelaar, Gen Re, Köln, 19 February 2010.
[21] Interview with a senior expert of Swiss Re, Zurich, 4 June 2010.

clear: develop two new global, generic, non-proprietary and public standards, i.e. a simple aggregated data spreadsheet standard on the one hand, and a detailed exposure standard that can also be used, for instance, for the binding procedures on exposure reporting in the United States. Both standards deal with the format and the content of data exchanged; they include Cresta codes, which now, however, represent only a very small portion of the specifications included.[22] Following a presentation by ACORD people on their plans during a 2007 Cresta meeting, a senior expert in charge of reinsurance information management did not mince his words: 'ACORD has emerged as the winning organisation for standardisation in the world, even if it remains largely American with around 50 out of 60 people based in the US. Cresta is really small and limited in comparison.'[23] This time, standards-setting developments have clearly been more successful. Both standards for an XML structured representation and a formalised spreadsheet of catastrophic exposure data used in the global reinsurance industry were published in November 2013. While Cresta codes are still used for the geographic entity of the aggregated exposure data, ACORD offers the guidelines for the format in which to exchange data.[24]

This successful outcome results not only from an evolution of the market of risk modelling that has become more mature and in which the largest reinsurers took back the initiative. The intrinsic nature of the standards under development also explains to a large extent why the move was more successful this time. While set within a strict private framework, the standard partially distances itself from the private extremity of the institutional continuum of standard-setting processes. It uses, indeed, a non-proprietary open source format. The standard is thus publicly available and provides a solution to convert multiple formats into a single interoperable instrument within the reach of all players of the value chain (risk modellers, (re)insurers, brokers, regulators). Regarding the material continuum targeted by the new standard, it does not merely set physical specifications for geographical zoning on a more detailed basis. It also includes more sophisticated data, whose content

[22] Interview with Juergen Heck, ACORD, Zurich, 4 June 2010. See also: 'ACORD Standards Working Group Ext. Request [for Catastrophe Exposure Data Standards]', internal document, n.d. [2009].

[23] Address by a senior expert, Cresta meeting, Rüschlikon (Zürich), 23 June 2007.

[24] See ACORD's and Cresta's webpages for further detail. Additional projects to develop standards for accounting and settlement communication processes are taking place within a new platform for eAccounting for the Global Insurance Market, the so-called Rüschlikon Initiative, named after the town on the outskirts of Zurich where Swiss Re has its large conference facilities. See the following LinkedIn page: www.linkedin.com/grp/home?gid=8129297.

regarding the nature and the value of insured objects is more substantial and of higher quality (such as the material used for the construction of an insured building, whether it is residential, commercial, or industrial, the type of its contents, the policy coverage terms, etc.). The standard thus addresses a slightly larger segment of the material continuum, with some extension towards the societal pole. To use the previously quoted wording of du Tertre (Du Tertre, 2008: 70–71), this new standard must factor in – even sketchily – the 'social relation of accessibility', i.e. the socio-historic and institutional constructs without which those batches of information remain meaningless and useless for any prospective unification of the natural catastrophe reinsurance market. Finally, as far as the spatial plane is concerned, the new standard is set in such a way as to have a global scope. To this end, however, it must take local and national specificities into account. This is why, for instance, it allows for reporting all necessary details for US property risks according to ACORD's ER3001 and ER3003 standards whose certification is required for delegated authority procedures concerning natural catastrophes in the United Sates. As Peter Hausmann, Head Cat Perils Europe Hub at Swiss Re and Co-Chair for data standards for the joint ACORD/Cresta NatCat working group, points out, 'detailed and accurate data is really required in the United States for the reason that there is a higher risk of litigation than in Europe, where a lower aggregated level is sufficient'.[25] This provides a genuine opportunity for non-American reinsurers (except US so-called captives in Bermuda) to improve their access to the United States, which for long remained at the margin of their standardisation's efforts.

Our journey so far has provided evidence that the internationalisation of insurance services relies only marginally on technical standards. This substantiates our argument that conditions for standardisation and internationalisation of service activities should not be viewed too restrictively as dependent on sectorial and institutional specificity. Should that be the case, the (re)insurance industry would be much more at ease with standardisation, as it is far from the ideal type of relational, non-material services oriented towards end consumers and relying on high-intensity labour – those so-called typical services resisting standardisation according to a restrictive sector-specific hypothesis. In contrast, cases studied here suggest that setting market standards for the insurance industry remains very difficult, and the few successful outcomes took place only after several abortive attempts. Established standards remain

[25] Interview with Peter Hausman, Swiss Re, Zurich, 4 June 2010.

essentially private and oligopolistic, narrowly technical, and deeply enmeshed in the logic of transnational markets. Some of the most recent developments suggest, however, some repositioning towards the centre of the standardisation axes of my topology, if only to include some public, societal, and endogenous dimensions of territorial sovereignty at the margins. Another type of standardised data exchange that is not limited to a defined branch of the insurance industry provides further evidence in this regard. My enquiry continues with those standardised guidelines for extra-financial reporting used by the largest listed insurance and reinsurance companies around the world.

Reporting Sustainably

While the history and current developments of exposure data exchange standards for natural hazards clearly belong to a larger trend supporting the globalisation of narrowly defined market-based instruments, insurers and re-insurers also use other tools, less strictly oriented towards the physical attributes of insured risks, and based more on historically and socially constructed values of how the risks are insured. Above all, insurers and reinsurers, as in other industries, make increasing use of reporting guidelines that aim to make large enterprises' social and environmental impact more transparent. The following analysis first provides some background on the emergence of sustainable reporting standards. Then, it presents the results of a systematic inventory of their use in the insurance industry, with particular focus on reinsurance and the Global Reporting Initiative (GRI) guidelines. The significance of exposure data exchange for the reinsurance industry (in particular for natural hazards) as compared to simple insurance and the fact that GRI is largely considered to be the leading sustainable reporting tool at the global level explain this specific focus.

The expansion of financial capitalism has prompted investors to ask for ever more detailed information regarding the financial health of companies in which they decide to invest. The spread of corporate social responsibility and environmental concerns has, however, encouraged companies to report information beyond narrowly defined financial performances. Social, environmental, and economic activities and related mitigating measures are increasingly conceived as integral to the information a company is expected to provide. Sustainable, societal, or extra-financial reporting describes an ordered publication of information on how a company appraises the economic, environmental, and societal impacts of its activities (Capron and Quairel-Lanoizelée, 2007). Beside ethical charters, codes of conduct, social certifications, and other evaluations by specialised rating agencies, sustainable reporting belongs to the new generation of management instruments in corporate social responsibility, whose origin goes back to the early 1970s and related struggles on profit distribution between employees, managers, and shareholders (Aggeri and Acquier, 2008). At the time, sustainable reporting lacked standardised formats and was focused on a limited number of issues. The growing importance of environmental issues in the 1980s and the rise of labour and human rights concerns in global value chains, has called for a much larger scope of sustainable reporting, together with the first environmental reports published separately from the yearly financial reports. While sustainable reporting has expanded to near normalcy, its concrete practices vary to a great extent and still remain largely voluntary. A number of benchmarks compete on the market and try to match a variety of institutional environments. Legislation differs widely both in the nature and meaning of sustainable or societal reporting; it can be more or less constraining (for instance, more for companies listed in Europe and Japan, less for those listed in the United States or elsewhere). The law can support and lead to legal action, particularly in the liability-based system of American common law. Moreover, the need for large listed companies to be accountable towards investors and civil society varies between countries and regions. More generally, the way corporate social responsibility is rooted in culture makes their practices differ greatly between the United States and Europe (Allouche et al., 2004; Acquier and Aggeri, 2008; Tsutsui and Lim, 2015).

Although we should avoid over-generalisations, differences between the United States and Europe are basically the following. In the United States, religious underpinnings make corporate social responsibility close to a moral charity exercise, rather than an institutional embedding of corporate conduct, as is the case in the European Union. The importance given to individual responsibility and freedom in the United States also leads to the view that it is up to the individual himself to act ethically, without legal strings that could impinge on his freedom. In contrast, Europe places more emphasis on collective and legal responsibility, with the individual regarded as an integral factor of the social fabric. Finally, the perception of the common good in the United States results from the ability of individuals to form a community by themselves, whereas, on the other side of the Atlantic, it explicitly results from a political construct. This in turn leads to an approach targeted at mitigating individual faults by charitable action in the United States, in contrast to a European approach focused on preventing potentially negative impacts of all sorts of corporate activities, in other words, by encouraging responsible action conceived directly at the core of the production processes of the enterprise.

Against the backdrop of such variations among existing practices, the development of a global standard can easily be seen as a handy way to reinforce the credibility of social and environmental reporting procedures and their comparison among companies – and even an absolute prerequisite for the benchmarking that supports funding and sales deals for companies. In less than a decade after their launch in 1997, the guidelines of the Global Reporting Initiative (GRI) became the gold standard for sustainable reporting. Although initially formed by two USA-based non-governmental organisations (CERES – *Coalition for Environmentally Responsible Economies* – and the Tellus Institute), with additional support from the United Nations Environment Program (UNEP), GRI has been an independent organisation since 2001, the year before it relocated its headquarters to Amsterdam. The initial aim focused on environmental reporting, but the scope quickly expanded to social reporting. GRI draws up guidelines that companies can follow for their social reports, using a harmonised format and an array of quantifiable social, economic, and environmental indicators. After several initial versions (G1, G2, G3, G4), it released so-called GRI Standards in 2016 as a consolidated set including all the main concepts and disclosures from the previous guideline, enhanced with a more flexible structure, clearer requirements, and simpler language. Specific supplements target certain sectors, such as finance or occupational health and safety.

GRI is the world's leading voluntary corporate non-financial reporting scheme. It is not only what KPMG describes as 'the most widely used voluntary reporting framework, far exceeding the use of national standards and other guidelines', with over three-quarters of the 100 largest companies in the world using it in one way or another (KPMG International, 2013: 31). GRI also greatly benefits from the importance of corporate sustainable reporting being explicitly referred to in multilateral diplomacy, official documents of the United Nations, new Directives of the European Union, and more or less constraining mandatory use at domestic and state level. For instance, GRI successfully launched an intense lobbying campaign before and during the Rio +20 Conference in 2012. This led to the inclusion of the following statement in paragraph 47 of the *Future We Want* Resolution adopted by the UN General Assembly: 'We acknowledge the importance of corporate sustainability reporting ... encourage industry, interested governments and relevant stakeholders, with the support of the United Nations system, as appropriate, to develop models for best practice and facilitate action for the integration of sustainability reporting, taking into account experiences from already existing frameworks.'[26] As the most widely used existing framework remains the GRI guidelines, GRI clearly has much to gain from such high profile intergovernmental recognition. Similarly, non-financial reporting is mandatory in the European Union for large companies (over 500 employees) both at the single and consolidated level, with GRI explicitly referred to among the various methodologies to be used to provide this information.[27] I will not discuss the effectiveness of GRI in general, as it lies beyond the scope of our enquiry focused on reference made to the instrument among insurance and reinsurance companies. Suffice it here to note conclusions drawn by scholars having studied in depth compliance issues in implementing GRI guidelines. According to Lim and Tsutsui, the use of GRI guidelines clearly follow a North–South divide; while developing countries are generally constrained to a substantive commitment, 'ceremonial' commitment drives the pack in developed countries, where by far most large listed companies are located and report their non-financial information (Lim and Tsutsui, 2012). In the same vein, Dingwerth and Eichinger point out that the relationship between transparency and empowerment supposedly reinforced by GRI disclosure approach is more conflictual than usually believed (Dingwerth and Eichinger, 2010).

The following analysis presents the results of a systematic inventory of the use of GRI guidelines in the (re)insurance industry by its largest multinational companies.[28] The results are summarised in Table 6.1.

[26] UN General Assembly, Sixty-Sixth Session, Resolution adopted by the General Assembly on 27 July 2012, A/RES/66/288.

[27] European Union, Directive 2014/95/EU of the European Parliament and of the Council of 22 October 2014 amending Directive 2013/34/EU as regards disclosure of non-financial and diversity information by certain large undertakings and groups. GRI is mentioned at paragraph 9 of the Directive.

[28] The analysis is based on data included in annual reports published by companies for the year 2014 (or 2013 if unavailable). For detailed results based on data collected in 2009–2010, see the following document prepared by Maude Gex, whom I warmly thank here for her extensive research: *The Use of the Global Reporting Initiative (GRI) Guidelines for Social Reporting: The Case of the Insurance/Reinsurance Sector* (mimeo, University of Lausanne, February 2010). Data update for 2013 and 2014 was undertaken in July 2015 with the help of Lucien Pamingle and Pierre-Alain Blanc, whom I thank here too. The empirical research was carried out on the social responsibility reporting of 34 of the largest multinational insurance and reinsurance companies: 24 insurance companies and the 10 reinsurance companies, located in 11 countries – USA (8 companies), United Kingdom (5), Bermuda (4), Germany (3), France (3), Switzerland (3), Canada (2), Netherlands (2), Japan (1), Italy (1), and China (1). The sampling of these companies is based on the 'diversified insurance' and 'life & health insurance' sections of the Forbes ranking of the 2000 world's biggest companies (April 2015 figures, based on sales, profits, assets, and market value).

Table 6.1 *Reporting practices of 34 amongst the biggest insurance and reinsurance multinational companies*

Rank	Forbes rank	Company	Type	Head	Separate	Gri	Version	Sector	Index	Level	External	Other
1	21	Allianz	a	de	Yes	Yes	g31	fs	Yes	a+	Yes	ungc, oecd
2	29	AXA	a	fr	Yes	Yes	ref	fs	No	Undeclared	Yes	ungc, iso, ifc
3	37	China life insurance	a	cn	Yes	Yes	g31	Not Used		Undeclared	No	No
4	42	AIG	a	us	Yes	No						No
5	50	Metlife	a	us	Yes	Yes	g4	fs	Yes	Core	No	No
6	56	Prudential	a	gb	Yes	No					No	Own
7	78	Zurich Insurance Group	a	ch	Yes	No						ungc
8	92	ING	a	nl	Yes	Yes	g4	fs	Yes	Comprehensive	Yes	ungc, iso, oecd, ifc
9	101	Munich Re	r	de	Yes	Yes	g4	fs	Yes	Core	Yes	ungc
10	128	Manulife Financial	a	ca	Yes	No						No
11	129	Generali	a	it	Yes	Yes	g4	fs	Yes	Core	Yes	ungc, iso, oecd
12	146	Swiss Re	r	ch	Yes	Yes	g31	fs	Yes		Yes	ungc
13	157	Prudential Financial	a	us	Yes	Yes	ref				No	No
14	162	Aviva	a	gb	Yes	No						ungc
15	175	Legal & General Group	a	gb	Yes	Yes	g4	fs	Yes	Core	No	ungc
16	179	Travelers Cos	a	us		No						No
17	181	Tokio Marine Holdings	a	jp	Yes	Yes	g31	No	Yes	b+	Yes	ungc, iso, oecd
18	194	Allstate	a	us	Yes	Yes	g4	Not Used		Core	Yes	
19	210	Lowe's	a	us	Yes	Yes	g4	No	Yes	Undeclared	No	No
20	216	ACE Group	a	ch	Yes	No						No
21	230	Aegon	a	nl	Yes	Yes	g4	fs	Yes	Comprehensive	Yes	No
22	276	Sun Life Financial	a	ca	Yes	Yes	g31	fs		c	No	No
23	286	CNP Assurances	a	fr	Yes	Yes	ref	No	No		No	No
24	328	Old Mutual	a	za	Yes	Yes	g31	No	Yes	c	No	No
25	372	Hartford Financial Services	a	us	Yes	Yes	g31	fs		c	No	No

26	461	Aon	a	gb	Yes	No						No
27	755	SCOR	r	fr	Yes	No						ungc
28	947	Everest Re	r	bm	Yes	No						No
29	1056	Partner Re	r	bm	Yes	No						No
30	1081	XL Group	r	ie	Yes	No						No
31	1164	Arch Capital Group	r	bm	Yes	No						No
32	1403	Axis Capital Holdings	r	bm	Yes	No						No
33		White Mountains Insurance Group	r	us	Yes	No						No
34		Hannover Re	r	de	Yes	Yes	g31	fs	Yes	b	No	ungc, oecd

Sources: GRI's Sustainability Disclosure Database; 2014 (or 2013 if unavailable) annual and separate CSR reports, websites, and other relevant documents pertaining to the CSR reporting of companies.

Explanatory notes:

- Insurance (a) and reinsurance (r) companies are ranked in the order of the Forbes 2000 ranking for the 'diversified insurance' and 'life & health insurance' sections combined (April 2015 figures, based on sales, profits, assets, and market value). The last eight companies are included in order to have a minimum of ten reinsurance companies in the sample and therefore have a base of comparison between insurance and reinsurance companies; their Forbes rankings are therefore non-consecutive.
- Head: headquarters' country according to ISO 3166–1 alpha-2 code.
- Separate: indicates whether the company publishes a separate report devoted to sustainable reporting. In the case of reporting included in the annual report, data is coded 'no'. Data is coded – in the case of no or minimal reporting.
- GRI-related information: A column indicates whether the company uses GRI guidelines; other columns indicate the version used (version of GRI (g3; g3.1; g4), whether they make use of their tailored sectorial supplement for financial services (sector; fs) required for a declaration at the highest application level, as well as of the index provided for communicating which items of the GRI disclosure have been reported (index). It furthermore indicates the application level, with levels ranging from C to A+ (highest) for versions 3.1. of the guidelines to determine the type and number of indicators reported and the new distinction between 'core' and 'comprehensive' for the G4 version (the 'core' option designs a report in accordance to essential elements of sustainability and the 'comprehensive' one designs additional disclosure and complete compliance with the guidelines). 'Ref' indicates sustainability/integrated reports that make explicit reference to being based on the GRI Guidelines, but for which there is no indication of the level used and no GRI Content Index. The column 'external' indicates whether the sustainability report is audited by an external third-party aggregated firm.
- Other: this column provides information regarding explicit use of other sustainability tools or standards in the annual or separate CSR report, with particular focus on the Global Compact of the United Nations (ungc), the Guidelines for Multinational Enterprises of the Organisation for Economic Cooperation and Development (oecd), the environmental management system ISO 14001 of the International Organization for Standardization (iso), and the Performance standards of the International Finance Corporation – the funding arm for the private sector of the World Bank group (ifc).

All thirty-four companies in the top layer of the global ranking publish separate reports with more or less detailed information on corporate social responsibility, even the reinsurance companies at the bottom of the ranking located in Bermuda, the specialised offshore financial centre for insurance. There are, of course, variations in the form of sustainable reporting, its content, and its methodology. But at least nineteen companies explicitly use the GRI guidelines, generally the latest version (G3.1 or G4, depending of the reporting year). Other tools used for social reporting include the UN Global Compact ten principles in the areas of human rights, labour, environment, and anti-corruption, the OECD Guidelines for Multinational Enterprises, the ISO 26000 guidance on social responsibility, and the International Finance Corporation's Environmental and Social Performance Standards. But these are far less commonly used than the GRI guidelines, and when they are, it is usually in conjunction with them. The GRI guidelines are thus a standard used by more than half of the companies included in the sample and by most of those who publish substantial social responsibility reporting. It stands out clearly as the most used standard compared to all others. Moreover, most of those using the GRI guidelines do it at the highest level, with the index provided for communicating which items of the GRI disclosure have been reported, and making use of the tailored supplement for financial services required for a declaration at the highest application level. Among those companies who do not have recourse to the GRI guidelines, we find a much greater proportion of reinsurers, in particular, all except one of those located in an offshore financial centre. Besides Bermuda, most of the companies that do not use the GRI guidelines are located in North America or the United Kingdom. Moreover, except for AIG, Prudential, and Zurich, none of them belong to the top 100 biggest companies, over three-quarters of which use GRI guidelines across all industries.[29]

It should be noted, however, that the transnational plane on which the largest (re)insurance companies make use of the standardised sustainable reporting practices of the GRI guidelines should not be seen as an undeviating social force. There are significant national and regional differences in companies' reporting practices. Companies with their head office in North America are less likely to produce such reports than those based in Europe and, if they do, do so differently. Moreover, Bermuda stands apart as the main offshore financial centre used by reinsurance

[29] In the aftermath of the crisis that saw AIG bailed out for the highest amount ever paid by the American taxpayer, AIG has been identified as the worst CSR company in the last five years by Corporate Governance and Sustainability International Group. See: http://cgsig.net/live_reports/aig-the-worst-company-in-csr-in-the-last-five-years/, accessed 27 July 2015.

companies. A continuum can thus be discerned from Bermuda and the United States, through Canada and the United Kingdom, to mainland Europe and Japan, with a marked difference between the United States and Europe. Roughly, the tendency in Bermuda and United States is for minimal, informal reporting, centring on local community involvement, charity, equality of opportunity, and sometimes the environment, all without using the GRI, in short, documents with few hard figures and little formal structure, perhaps even reduced to a list of contributions to charity. At the other end of the spectrum, the tendency in mainland Europe and Japan is towards more comprehensive and structured reporting, addressing the three pillars of sustainable development (economic, social, and environmental), providing both quantitative and qualitative information on the social and environmental impact of the company's activities and on measures planned or already in place to reduce their negative impact, with much more frequent and precise use of the GRI. Of the ten companies using GRI directly, seven are European and one Japanese. Canada seems to sit in between the two extremes, while UK companies adopt the same approach as mainland European companies, but without using GRI.

Among the factors explaining the regional and national variations in the use of standards for social reporting practices, national regulatory and socio-economic environments, together with differences in approach to CSR and to climate change, are among the most significant. While evidence gathered so far has lead us to question too sector-specific an explanation of the role played by standards in the internationalisation of services, this suggests that – at least in the domain of sustainable reporting – an institution-specific explanation is likely to make sense. In fact, this inventory suggests that standards used by insurers and reinsurers are not strictly limited to a small number of private and oligopolistic, narrowly defined technical specifications sought for accessing markets in an exogenous logic, supposedly valid on a worldwide basis. The extensive use of GRI guidelines suggests that – as for many other large listed companies that are sensitive to their social and environmental reputation – they also include standards less tightly confined to physical attributes and the private sphere. The information supposedly guaranteed by the standard makes the socio-historical underpinning of the quality and security requirements more visible. The regulatory and broader socio-economic environments, together with opposing approaches of corporate social responsibility, go a long way towards explaining regional and national variations in the use of standardised instruments of sustainable reporting.

* * *

The story advanced in this chapter contributes to building the case made throughout this book. Insurance standards, as for other service industries, should be explained from an extensive understanding of the power plays and conflicting political economy objectives set in motion by designing quality and security attributes required in the economic transaction. Such an extensive hypothesis posits that standards, if and when set, link national economies to global markets by fixing quality and security uncertainty in various ways. With an in-depth study of how the (re)insurance industry relies on standards to create new markets and preside over changes in their performance, this chapter has provided some ground to engage this argument. As my examination has shown, the (re)insurance industry fiercely struggled to set standards, with many abortive attempts and finally a few successful ones. This finding clearly differs from conventional views that consider standardisation in a non-typical service industry such as (re) insurance rather likely, since it is non-relational and tangible, strongly oriented towards a business clientele, and more capital- than labour-intensive. The development of standards that help to shape new (re)insurance markets does not reflect this restrictive sector-specific hypothesis. Indeed, it appears to be less dependent on intrinsic attributes of the industry.

Among the few established standards or those under development, the ones for calculating longevity risks traded in highly securitised life markets, for generic data exchange formats, and for accumulation controls of risk exposure to natural catastrophes are essentially private and oligopolistic, narrowly technical, and reliant on an exogenous compliance system associated to a transnational market logic. Interestingly, more recent initiatives show some evolution along larger segments of my standardisation typology, and therefore away from mere private, physical, and transnational extremities of the framework.

In the field of life insurance and the standardisation of securitised longevity risk, the potential role of governments in longevity bonds, the continuing relevance of individual judgement and customised contract, and the lack of really global markets for such niche securitised instruments suggest indeed some distantiation from those extremities. While standardised contracts and pricing provide liquidity as for any other financial market, they also provide substantial means for an effective hedge of the risk borne by bonds issued on the new 'life market'. As hedge effectiveness competes with liquidity requirements, those actors most likely to win on the new 'life market' agreed that standardisation was not only essential but should also be set within this trade-off. A similar configuration characterises the evolution of standardised formats used in data exchange at the core of a proper functioning of the reinsurance industry. While still overly private, narrowly technical, and fairly transnationalised, we are beginning to see a more complex picture. Standards in some way embrace a public dimension, if only because discussions on the overall architecture for generic data exchange take place within the framework of the Centre for Trade Facilitation and Electronic Business (UN/CEFACT), itself an instance of the United Nations Economic Commission for Europe (ECE). Moreover, the new ACORD standard for accumulation control of risks of natural hazards successfully managed to bring the trio of model providers on-board for an open source and non-proprietary format. Furthermore, not content with an enhanced granularity of Cresta geographical zoning, the standard now includes more information directly related to societal values, such as the content of the insured object or the coverage terms of the insurance policy. In a different domain but with a similar logic, the rapid surge of sustainable reporting among (re)insurers, as with other large companies, offers compelling evidence about how standardised information and data exchange move away from narrow financial reporting to include a number of calibrated social and environmental values (non-financial reporting). Finally, as far as the recognition of authority of standards is concerned, the development of standards for accumulation control of risks of natural hazards since its heroic beginnings in the 1970s can also be seen as the loosening of the quasi-imperial domination of the Munich Re/Swiss Re duopoly. Whereas the two giant reinsurers largely imposed their views on the matter, first in Latin America, and later on in Europe, the global reach of the new model providers and the weight of the United States in the experience acquired by ACORD compelled the duopoly to design a format that would include the certification procedures in place in the United States and more equitably recognise worldwide natural catastrophes' specificities and insurance policies' coverage terms. We saw for instance that that the ACORD standard is fully compatible with US ER3001 and ER3003 standards whose certification is required for delegated authority procedures concerning natural catastrophes in the United Sates.

In all these domains, the international standardisation of insurance services faces significant obstacles. Although far from easy and yet applicable to small numbers, it brings into play an ambiguous form of authority that is neither private nor public. The way in which it expects to lower the uncertainty of market transactions rests on quality and security requirements that are neither completely physical nor exclusively societal. And last but not least, we have seen throughout this chapter that the compliance expected to those standards is primarily based on the market logic of an ever more transnationalised and financialised capitalism, exogenous to the territorial space of state sovereignty. Be that as it may, the authority of those standards is recognised on the endogenous basis of the exclusive conception of territorial sovereignty. If only at the marginal level, the standard itself includes provisions in conformity with certification and accreditation procedures of some of the most important national jurisdictions. In sum, insurance standards reflect a truly transnational hybrid authority. Far from being set only by powerful actors in order to homogenise narrowly defined technical specifications that would support a higher level of market convergence across countries, the common understanding of the quality and security of insurance services encapsulates a public dimension that tends to blur the distinction between the private and public spheres, and their scope cannot ignore societal values as well as overlapping with the supposedly exclusive sovereign spaces of territorial states.

7 The World Office

Standards and Business Process Outsourcing in India

In Chapter 3, we saw how the offshoring of services has become a powerful and significant phenomenon in contemporary capitalism. The shift began in the 1980s with outsourcing contracts in data processing and call centres at the bottom of the value chain. Far more advanced sectors now include legal, fiscal, medical, architecture, consulting activities, and many sorts of business services enabled by information technology. In the same chapter our discussion of the drivers of service offshoring stressed the importance of looking beyond ICT, labour costs, and the mobility of service providers and consumers to take due account of language and cognitive skills, cultural understanding, and various kinds of geographical links likely to support the tradability of services. The scope of industries concerned goes beyond conventional views on barriers to the tradability of services focused on the specificity of distinct service industries and institutional environments shaped by government policies usually referred to as non-tariff measures. In contrast to restrictive hypotheses on the standardisation and internationalisation of services confined to distinct industries and their relations to national institutions, my analysis emphasises an *extensive hypothesis*: service standards can link national economies to the global marketplace by responding to quality and security uncertainties that can accommodate opposing political economy objectives and power configurations. It is from this perspective that the present book analyses service standards as a form of transnational hybrid authority whose scope extends from physical measures to societal values, blurs the distinction between private and public actors, and reinforces the deterritorialisation of regulatory practices in contemporary capitalism.

In contrast to prior chapters on (re)insurance, in this chapter, we focus on activities that match more closely the ideal type of a relational, non-material service, relying on high-intensity labour, some of which is oriented towards the end consumer – in short, services understood in conventional accounts as less likely to be standardised and internationalised. We examine more specifically the offshoring of business services.

While labour intensive, such activities are supported by a great deal of information and communication technologies; they are thus often called IT-enabled services (ITeS), or business process outsourcing (BPO) when the specifics of the tasks outsourced are seen as particularly important. More than any other country, India epitomises the extent to which the expansion of such segments of the tertiary sector is likely to follow the rise of the global knowledge-based economy. Despite the rise of alternative locations in the Philippines, Eastern and Southern Africa, the Maghreb, Latin America, and Eastern Europe, despite the 'death of outsourcing' announced by the giant consulting firm KPMG to which we referred at the beginning of Chapter 3, India remains the world's leading outsourcing location. Throughout the post-2008 global economic crisis, India not only continued its double-digit growth of the industry; it even increased its market share in the global sourcing industry. According to the National Association of Software and Services Companies (Nasscom), the voice of the IT service industry, India accounted for 55 per cent of the global outsourcing market in 2017 compared with 52 per cent in 2012 and 51 per cent in 2009 (Nasscom, 2012, 2018). In a keynote address at an India–China business forum in 2015, the Indian Prime Minister, Narendra Modi, did not hesitate to use a cliché to compare the two giant emerging economies: 'You are the "factory of the world"; whereas, we are the "Back office of the world". You give thrust on production of hardware, while India focuses on software and services.'[1]

Being identified as the back-office of the world indisputably endorses India's achievement over the last two decades; however, it also recalls the undemanding and repetitive tasks performed by low-skilled and cheap labour in call and customer centres – the archetype of jobs in an industry described by its critics as electronic sweatshops (Garson, 1988), panopticons of the workplace (Fernie and Metcalf, 1998), or assembly lines in the head (Taylor and Bain, 1999). Considering the range of services performed in India, PM Modi's cliché might thus look surprisingly self-defeating. For almost two decades, parts of the industry have moved away from basic back-office tasks such as outbound calls for marketing anything and everything from insurance to dodgy pills, inbound calls dispatched to remote customer centres, data processing, and software coding. These days, many companies operating in India can rightly claim to be part of a fully-fledged office of the world, not just undervalued peons. Young Indian PhDs prepare patent profiles for new drugs

[1] 'Read full text: Keynote address by PM at India-China business forum', *The Times of India*, 16 May 2015, online at: http://timesofindia.indiatimes.com/india/Read-full-text-Keynote-address-by-PM-at-India-China-business-forum/articleshow/47304933.cms.

developed by Indian, European, or American drug companies; other graduates develop complex financial products for major investment banks in London, New York, or Zurich; lawyers draft case briefs for giant law firms in the United Kingdom and elsewhere; actuaries develop models to assess risks in diverse insurance lines; engineers design key components of the next generation of aircraft and write software for their flight control.[2] Entrepreneurs have even rebranded the industry to move it away from what they see as an outmoded name, outsourcing. In 2012, Nasscom began a systematic rebranding strategy by referring to the industry as business process management (BPM) to emphasise the transformation experienced since its inception in the early 1990s; by doing increasingly complex work, performed in India or elsewhere nearer the clients by foreign affiliates of Indian companies, seeing itself more as a partner to its clients than a mere subcontractor, the industry should thus be better identified as 'a full-service value provider rather than an industry that plays only in the lower-end of the services spectrum … [and] give India a seat on the high table'.[3]

In this chapter, I build on the wide range of activities provided by the Indian office of the world to revise the conventional account that intangible and relational services are harder to internationalise and standardise than so-called industrialised services. The Indian office of the world, whether a disguised electronic sweatshop or an intrinsic partner of its customers, must codify the disaggregation of service production and delivery into discrete processes likely to be assessed against distinct quality performance and security guarantees. A wide range of activities and institutions have shaped the development of the Indian service industry, including a comprehensive use of standards. In examining the ambiguous transnational hybrid authority exerted by such standards, I focus on the three analytical dimensions used earlier in this book. The first is the extent to which such standards are set by actors able to bridge the public and private spheres; the second considers that even highly technical and managerial concerns cannot ignore social and cultural values; finally, the third looks at how standards require recognition from both transnational market forces and the territorial state.

I begin with some background on how India became the world's office. In contrast to conventional views confined to a state/market divide, I highlight that the development of IT services and business process outsourcing in India and their current and future challenges involve a

[2] Those examples are adapted from Sharma (2015: 185).

[3] Nasscom, 'From BPO to BPM', online at: www.nasscom.in/overview-9, accessed 12 July 2016.

complex relationship between global structural change induced by technological innovation and market constraints, foreign affiliates of multinational corporations, state policies, and local contexts. The chapter continues with an analysis of the rise and range of international standards and certified management tools used in business process outsourcing in India. Finally, the particular role of Nasscom, the voice of the Indian IT service industry, is considered in some detail, from the time when service offshoring began scaling up to more recent initiatives that attempt to transform India into an innovative standard maker. While conclusions that arise from evidence provided in the chapter can only be sketched out, the important point remains the ambiguous authority of service standards in India, where the institutional nature and political economy content of standards do not count for much, as long as they help to provide 'whatever the client asks for'.

India and the Not-So-Flat World of Services

In his best-selling account of the new 'flat world' of twenty-first-century globalisation, Thomas Friedman gives the Indian service industry a prime role, which resembles that of a dream business story,

> And so with Y2K [2000] bearing down on us, America and India started dating, and that relationship became a huge flattener, because it demonstrated to so many different businesses that the combination of the PC, the Internet, and fibre-optic cable had created the possibility of a whole new form of collaboration and horizontal value creation: outsourcing. Any service, call center, business support operation, or knowledge work that could be digitized could be sourced globally to the cheapest, smartest, most efficient provider (Friedman, 2006: 131–132).

This exciting view also introduced to the whole world – or at least to those millions of readers of the New York Times columnist – the daring part allegedly played in the story by Nandan Nilekani, the founder of Infosys Technologies, 'one of the jewels of the Indian information technology world', whose global conferencing centre in Bangalore is described as 'ground zero of the Indian outsourcing industry' (Friedman, 2006: 5,6).

Studies portraying the success story of the Indian service industry and its prominence in the global market for outsourced services have proliferated over the last two decades. What makes India's position among large emerging powers so distinct, in particular as compared to China's strategy based on mass manufacturing, continues to be widely debated. Beyond the entrepreneurial skills of the handful of captains of Indian industry popularised by best-selling books on management, most analyses remain stuck in a narrow-minded state/market divide. Liberals take the dramatic shift that came with the liberalisation policies adopted in the early 1990s as a turning point – with some discussion on the respective weight of internal or external pressures (Heeks, 1996; Nayyar, 2012: 48ff). Some studies take the opposite view and focus on the role of the developmental state in technological innovation for late industrialised economies, in which India's success story in services echoes its well-crafted policies to build industrial capabilities in the pharmaceutical and health industries (Saraswati, 2008; Sarma and Krishna, 2010).

In contrast to both those narratives, the account provided in this book shows that the development of IT services and business process outsourcing in India build upon a more complex relationship between global structural change, foreign-affiliates of multinational corporations, state policies, and local context. In a much-quoted article, Dossani and Kenny set out to explain the dynamics of offshoring 'from the perspective of the firm, the industry, and the recipient country' (Dossani and Kenney, 2007: 773). However, even in that perspective the authors stress that the developments that made Indian service offshoring feasible depended not only on multinational firms and some early Indian entrant firms but also on a wider span of market institutions: 'The growth in offshoring is intimately linked to the prior development of India's software sector and an enabling regulatory and other institutional environment' (Dossani and Kenney, 2007: 773). More explicitly, Parthasarathy (2013b: 383) emphasises 'the need for a nuanced, evolutionary understanding of offshoring'. In this perspective, the State is important, but not any state can effectively play a development role (Evans, 1995; Parthasarathy, 2004). As Srinivas highlights in her analysis of technological advances and market regulation in health industries in India and other emerging economies, the time has come to give up frontal oppositions between states and markets: 'Markets are contingent constructions of specific moments in technological advances, not least because change occurs in particular places, not in the abstraction of nations'; from that standpoint, the 'fundamental challenges for nation-states are to wed technological advance to local institutional context, as well as international standardization pressures' (Srinivas, 2012: 226). Based on such an evolutionary understanding, it is important to emphasise the processual, sequential, and overlapping dimensions of the wide range of institutions that have shaped the development of the Indian service industry and their comprehensive use of standards. After a first overview of the history of the Indian service industry, we will focus more specifically on the significance of technical standards among those different institutions either reinforcing or overcoming path-dependent advances of the industry.

How It All Began

As with many technological innovations, the early history of Indian IT services is closely linked to military technology and defence spending. The Indian nuclear and space research establishment began to invest in IT capabilities in the wake of independence in the 1950s. It depended on affiliates of multinational corporations (MNCs) to have access to hardware bundled with software. While most of the work was done within MNC affiliates, a few defence contracts started to outsource some software development to local firms – a move made possible by a very high standard of secondary and higher education in cities like Bangalore and Mumbai, where much of the civil service and defence industry was located after independence. As an OECD study on the growth of the Indian software industry points out, 'one of the biggest contributions that the public sector research establishment made to the Indian software industry was to provide a nucleus of highly skilled engineers and scientists' (OECD, 2000: 133). In 1971 the importance of this nexus between MNCs, state procurement policies, and a local legacy prompted the Indian government to establish the Department of Electronics in order to provide a stronger and more coordinated impetus to the industry. The so-called Software Export Scheme was adopted the following year to extend access to the required hardware. In due time, developments of the industry owed much to the particular role played by one multinational firm: International Business Machines Corporation, a.k.a. IBM.

IBM began operations in India as early as 1951, and the firm rapidly secured a quasi-monopoly in data processing machines and services. The originality of IBM's operations in India is that most of its revenues came from importing discarded machines from the American and European market, refurbishing them locally, and leasing them out to Indian users at very high rental prices. For instance, in 1975, computers used for those services cost around $1,200, while IBM in India charged $20,000 or more as annual rental for similar machines, with prices quoted in dollars for products and services manufactured in India – a covert practice in breach of Indian law (Sharma, 2015: 61). Concerned by such abusive practices and adamant that foreign direct investment should fall in line with developmental priorities, the Government of India and parliamentary committees paid particular attention to IBM's operations concerning balance of payment, domestic competition, labour markets, and technological innovation. After several failed attempts, the adoption of the Foreign Exchange Regulation Act in 1973 provided a legal mechanism to curtail the firm's abuses. By requiring foreign companies working in India with more than 40 per cent foreign equity to obtain fresh approval from the Reserve Bank of India to continue their operations, this legislation prompted an intense power struggle between the Government of India and IBM, which opposed minority Indian shareholding in its manufacturing, sales, maintenance, and other service operations. After much wrangling, IBM was asked to withdraw from India due to its unwillingness to comply with the rules on foreign exchange. In May 1978, it did so and ceased all operations until its re-entry in the mid-1980s, first in joint ventures with the Tata group, and later as a fully owned IBM affiliate since the late 1990s. In 2014, IBM India was active in every segment of the Indian IT market – hardware, software, research, business process outsourcing (BPO), and consulting. With some hundred thousand employees in the country, it portrayed itself as the largest foreign employer in India (Negandhi and Palia, 1988; Athreye, 2005; Sharma, 2015: 55–75).

The IBM story is important in the sense that it points up already existing capabilities in the Indian IT landscape. With policy explicitly designed by the Government of India to support the industry and a relatively abundant pool of skilled labour available, the emergence of local companies began in the late 1960s. Tata Consultancy Services (TCS) was established in 1968 as a division of Tata Sons, the largest Indian group active in wide-ranging activities in engineering, chemicals, consumer goods, and services (Ramadorai, 2011). Usually considered India's first software services company, TCS was also the first firm to export software in return for access to imported hardware in 1974 (Heeks, 1996: 69). With a view to further develop its software services, it is probably no coincidence that TCS created a joint venture with the American service firm Burroughs the same year as IBM left India. Tata Burroughs Ltd (TBL), as it was known at the time, started to export software services in 1978 and many other firms soon followed suit. Many firms that began with other operations moved to software services as their core business, developing customised software both on and off site (OECD, 2000: 134). Today, TCS remains the largest Indian IT service firm. Although a company was specifically created for maintaining computer systems after IBM's departure, in 1978, more than 1,000 programmers found themselves on the job market. Their best option was either a visa clearance to find a job in the United States or to create (or join) a small or medium enterprise in India. While the latter is typically praised by studies emphasising Indian entrepreneurship taking advantage of new market opportunities, the former fuelled the trend of 'body-shopping': service offshoring resting on genuine delocalisation of bodies, farming out Indian software professionals to the clients' sites to execute short-term projects. As shown in much detail in a study on how the American

IT industry invented this highly innovative global labour market management system, India thus became 'not only a source country of flexibilized IT labor, but also a coordinating center for global labor mobility' (Xiang, 2007: 10). At the turn of the millennium, probably more than one thousand agents were supplying as many as 20,000 temporary Indian IT workers across the United States (Xiang, 2007: 4).

Where Standards Come In

Standards played a less known, but significant, role in the unexpected consequences of the dramatic exit of IBM for the place of India in the world of services. Initially, the objective of the Government of India was to support the endogenous development of an IT hardware industry. However, faced with an extremely low rate of computerisation and a highly fragmented market, the standardisation of the hardware sold on the domestic market remained weak. This in turn discouraged firms from selling all-included packages of software bundled with their hardware, as, for instance, was the practice in Taiwan and Japan. Instead, they provided separate software services or none at all; this move prompted the emergence of small and independent local firms specialising in the development of ad-hoc software and in-house developments in larger companies. Thus, while the initial objective of the Government of India was to support the manufacturing of an indigenous hardware industry in order to increase access to computers, the lack of standards in that segment of the industry resulted in the unexpected emergence of IT services that would soon be ready for a huge surge in export markets (Saraswati, 2008: 1147; Niang, 2013: 240). Basically, the winding-up of IBM operations made room for the flurry of local software service companies created as a substitute for the lack of standards in the burgeoning Indian IT hardware industry.

It is against this backdrop that changes to the underlying interests in IT policy formation occurred years ahead of the conversion of the Indian economy via liberalisation reforms in the early 1990s. India's emergence as the prime low-cost destination for IT services offshoring began in the 1980s. While the initial industrial policy was driven by hardware protectionist interests, the growing ability to capitalise on the software industry's export potential led the Department of Electronics to change course. The bulk of the credit for this transformation usually goes to the technophile politician Rajiv Gandhi, who succeeded his mother, Indira, after her assassination in 1984, although she had initiated the policy shift during her second term, which had begun in 1980. Despite the continuation of the strong link established by Nehru between the

Indian state and prominent scientists, there was a shift away from a main focus on basic science to supporting state-led import substitution infrastructure. The new alliance was between technophile bureaucrats and private sector avant-garde technologists (Sharma, 2015: 99–103). The first major policy change was the Computer Software Export, Software Development and Training Policy of 1986, explicitly aimed at increasing India's share of world software production and gaining access to global technologies. In the words of N. Seshagiri, who took over as director of the Computer section in the Department of Electronics in January 1982, the basic means to achieve this was the 'flood in, flood out' method, i.e. allowing an initial flood in of imports to achieve a greater flood out of exports.[4] Overall, the significance of this policy rests on much easier access to imported software packages and many measures supporting the export of software services (Saraswati, 2008: 1148). The next measure that put the industry on a launching pad was the gradual development of several state-run Software Technology Parks (STPs) in the 1980s and the decision in 1991 to set up an umbrella body called Software Technology Parks of India (STPI) to run these parks as autonomous entities under the supervision of the Department of Electronics. STPI not only ensured forward-looking management of STPs, including guarantees regarding tax exemptions and financial incentives but also provided indispensable services, especially high-speed data links through satellite earth stations, to attract foreign multinational corporations and support burgeoning Indian firms; no less important were large exemptions granted for preferential access to land and for labour law holidays (Upadhya and Vasavi, 2008; Upadhya, 2009).

In 1991, the same year as the STPI was established and only days after the new Congress leader, Narasimha Rao, was sworn in as prime minister in the wake of the elections held after the assassination of Rajiv Gandhi, a balance of payment crisis forced the government to request financial assistance from the IMF. Together with the World Bank, a broad set of policy reforms was adopted with the aim of an immediate stabilisation programme and longer-term liberalisation by opening up the Indian economy to more competition both from within and abroad. Arguably, the abolition of the 'license-permit Raj' in July 1991 best epitomises the reform package that put an end to existing licenses for many business decisions, import and export practices that had existed since independence. The opening up of the economy undoubtedly had a major impact on the rise of Indian IT services on the global market. Yet,

[4] Dataquest, 'The New Software Policy: Dr. Seshagiri Clarifies', January 1987, pp. 82–95, quoted in: Parthasarathy (2013b: 385)

I have provided ample evidence of the importance of continuing state intervention in the transformation of the industry – a role that should be placed in a wider historical and institutional perspective, including the Indian education system fed by a dense network of Indian Institutes of Technology (IITs) and Indian Institutes of Management (IIMs) dating back to the 1950s. While foreign companies built up technology transfers and capital contribution, they also had the advantage of highly skilled local staff. As recalled by Raman Roy, known as the father of the BPO industry in India, the bet was 'to bring the Indian perspective and be confident enough of the greater efficiency of the Indian workforce'.[5] In the opening of this chapter I also brought to mind the prime importance given by Thomas Friedman's *Flat World* to the so-called millennium bug in putting Indian IT services on the world's map. The story of how the Indian IT industry left its footprint across the globe has now been told again and again. Less known, however, is that it was not too long before standards once more played an important role.

We Provide Whatever the Client Asks For!

From 1991 onwards, the more liberal environment of important parts of the Indian economy triggered the entry of an increasing number of foreign firms in IT-enabled services. Pioneer experiences involved American companies such as American Express, General Electric (GE), and Texas Instruments in operating backroom functions, or airlines such as Swissair and British Airways in handling accounting operations. Those so-called captive units of multinational banks and airlines in the early 1990s rapidly matured into a multibillion-dollar industry. The entry of foreign firms helped local Indian software firms to acquire the required expertise to meet global standards in an ever-wider range of IT-enabled services. Moreover, as export zones dedicated to the software industry, STPs provided world-class communication facilities, massive tax incentives, privileged access to land, and labour law holidays for offshore services. Thus, instead of having programmers and other types of IT-related service providers scattered at client sites across the world, the tide of 'body-shopping' could be reversed, as services were now provided directly from India (Parthasarathy, 2013a: 387). However, this posed new challenges to an industry which soon realised that cost alone (i.e. exploitation of skilled, globally mobile, and cheap labour) could not do the job for long. Quality concerns needed to be addressed as well.

[5] Interview with the author, Gurgaon, India, 8 February 2008.

With body-shopping, quality chiefly depended on the programmers sent overseas and the reliability of the placement agents. With services offshored directly from India, the whole business process was involved in ensuring that orders would be delivered according to plan and in time. Gaining recognised conformity to existing international standards on quality management systems would quickly be seen as the only way forward.

Quality standards thus provided crucial tools without which the offshoring of service activities previously located in developed countries would, for the most part, have been unlikely. They contributed to overcoming the conventional resistance to relocation conveyed by the assumption that only basic, repetitive tasks could be outsourced to offshore locations. As Dossani and Kenney (2007: 775) remind us, foreign investors in service offshoring drew heavily on the 1990s managerial culture of reengineering by decomposing and standardising all sorts of business practices: being able to prove your conformity to standards recognised in the industry helped 'business decision makers [to] be persuaded that offshoring was an acceptable strategy or 'legitimate' ... by proving that there were appropriate levels of security and sufficient assurances of business continuity. ... The point was to create the perception that moving one's service operations to India was not 'unusual' or 'risky', but rather was part of a normal business model.' On the other hand, for the new Indian IT service firms, standards could be used as a response to reputation threats driven by their nouveau-riche destiny. As Rajesh Kalra, Chief Editor at the Times of India Group, points out, 'BPO remains a sector where it's easy to make quick money in setting up new businesses without too much regarding on quality and standards'.[6]

As the number of actors in the industry grew and became more heterogeneous, with their reputation increasingly at risk, the acquisition of quality certifications belonged to what the industry often refers to as a key 'differentiation strategy' (Banerjee and Duflo, 2000; Athreye, 2005: 408). For the person in charge of quality and process excellence in one of the major Indian BPO firm, standards build trust in this context: 'in our industry, the product *is* the transaction; when you do a call, that's a product for us; product quality is thus embedded in business operations with a team designing a framework for quality; this is how process standards became important.'[7] BPO pioneers of India as well as the younger generation of quality managers in charge of operations share

[6] Interview with the author, Noida (New Delhi), 19 January 2016.
[7] Rajesh Sehgal, Head of Quality and Process Excellence, Wipro, interview with the author, Sarita Vihar (Delhi), 22 January 2016.

this view: Raman Roy, one of India's BPO pioneers, was himself associated with the early development of a standard specifically dedicated to customer centres; Sudeep Banerjee, former President for Enterprise Solutions at Wipro, one of the Indian big three ITeS/BPO company, equates standards with calling cards: 'Wipro could claim being able to implement all sorts of quality standards at world level, even if those standards were not written by us.'[8] Likewise, the person in charge of internal audit and compliance in one of the Indian BPO firms with the fastest growth in recent years insists that standards were 'very important, because every time you go requiring new clients, the first thing the client want to see is whether you are an established player; it's a kind of credibility which, initially, was just to be in the game, but now it's BAU [business as usual] for us, ingrained in the normal processes of the company.'[9] It is not surprising, then, that at the height of the rush for certification with worldwide recognition, the Indian IT and BPO industry was notoriously known for exhibiting the largest number of quality certifications achieved by any single country, with more than 50 per cent of all certificates in the most significant segments of the industry and more than a third of worldwide entities registered with what was seen as the industry's gold standard – CMMI level 5 (Nasscom, 2010: 185–186). With such a massive use of certified standards, the outsourcing of business services in India achieved within roughly a decade what a World Bank study already suggested in 1994 as a vital transition from the 'low cost, low quality' to the 'low cost, high quality' quadrant of the global market (Hanna, 1994: 246).

After having weathered the global financial crisis relatively unscathed, the service offshoring industry has positioned itself as beyond the labour arbitrage and differentiation strategy that marked the emergence of the Indian office of the world. The new concepts à la mode are 'verticals' and 'digital' (Nasscom, 2018). Verticals stand for an organisation strategy aimed at delivering end-to-end highly customised services within a wide range of sectors as a form of advisory partnership. In the words of mid-level management officials in one of the leading BPO Indian firm, 'vertical brings the depth to content, with skilled people on the domain concerned … it's a process excellence, in which we pretty much service the entire value chain of services provided'.[10] For its part, digital denotes

[8] Interview with the author, Bangalore, 12 February, 2008.
[9] Amit Sharma, VP internal audit & compliance, EXL services, interview with the author, Noida (New Delhi), 20 January 2016.
[10] Vineet Malhotry, Sr Director for marketing, Cognizant, Interview with the author, Gurgaon (New Delhi), 19 January 2016.

a quality and security of the service provided built upon platforms directly enabled by the automated and robotised service delivery systems. The industry began with some hidden programmers and cheap labour working thousands of miles away, and moved on to B2B and B2C contact centres with increasingly specialised and complex tasks. It now spreads across the new world of big data and robotics: loan application procedures, insurance claims documentation, health service book accounting, or service desk calls answering; these are just a few examples of the tasks that can be carried out by what is increasingly known as robotic process automation.[11] According to NASSCOM, India is now the 'hotbed for digital innovation with a rich eco-system of start-ups, tech providers and service providers engaging in global delivery' (Nasscom, 2018: 213). This could suggest that process standards could lose their importance altogether as they would be replaced by algorithms engineering fully digitalised platforms. Yet, when asked about this, mid-level management in charge of quality and security certifications unanimously give a negative answer. They assume that even at a later stage, when service organisations could have fully integrated robotics with artificial intelligence, it would only require inventing really different types of standards. As it is unlikely that the whole process of an outsourced business will be automated, there will always be a part that has to be managed and intermediated and therefore in need of a standard against which assessing quality expectations.

It is one thing to remember that quality standards have been instrumental in the ability of the BPO industry in India to claim that it can provide whatever the client asks for. It is quite another to explain what is standardised by whom and where. In order to do that, we must go beyond the assumption that standards matter. By turning our attention to the three what, who, and where questions guiding the enquiry of this book, my investigation aims to uncover the ambiguous juxtaposition of power instances supporting the transnational hybrid authority of standards.

What is Standardised?

What exactly are the standards used across the Indian service offshoring industry to disaggregate repeatable and measurable tasks on a reliable basis at the global level? Does their scope support my extensive hypothesis on their role in the globalisation of services beyond the comfort zone

[11] Andrew Burgess, 'How robotics is changing the face of Business Process Outsourcing', *Robohub News*, 7 January 2015, accessed online on 15 June 2016: http://robohub.org/how-robotics-is-changing-the-face-of-business-process-outsourcing.

of industries identified as the most easily tradable and least dependent of the national territorial framework within which service relations are institutionalised? In other words, examining in detail the types of standards most commonly used by the business services outsourcing industry in India allows us to consider to what extent they reflect an ambiguous juxtaposition of technical specifications with societal values, opposing political economy objectives, and power configurations.

Offshoring of services in India unquestionably rests on many narrow technical specifications related to IT and other aspects of the infrastructure used for that purpose. The question here is rather about the quality assured by the *processes* performed by the service providers in charge of the outsourced tasks. It is for that reason that standards in management systems and business processes have become integral components of service offshoring with a whole range of dedicated quality and security standards. ISO 9000 series are by far the best-known and most widely used in India.

Quality management standards were first developed in defence contracting in the United States and the United Kingdom during World War II and were later expanded by the British Standards Institution in order to address the growing internationalisation of production networks. The first ISO version was published in 1987, with successive revisions; the latest was adopted in 2015. More than a million certificates were issued worldwide in 2014, and India was the fifth country among them after China, Italy, Germany, and Japan; it remains the most popular quality management standard in the world.[12] As we will see, many other management systems and process standards are used in India. For the time being, let us bear in mind that such standards help to legitimise the transnational hybrid authority of non-conventional forms of market creation and regulation discussed in this book. As Tamm Hallström and other scholars have shown (Tamm Hallström, 2004; Higgins and Tamm Hallström, 2007; Gibbon and Henriksen, 2011), management systems standards do more than establish technical specifications designed to ensure quality. By objectifying, codifying, and reengineering management processes along a so-called Plan-Do-Check-Act cycle at all levels in the organisation, they convey a particular form of power that allows for an ambiguous mix of what can be standardised along the two poles of the physical and societal worlds. Indeed, in spite of not targeting the size of bolts and nuts, system management and process standards nevertheless span the material continuum of standardisation. Conformity assessment

[12] The ISO Survey of Management System Standard Certifications – 2014, online: www.iso.org/iso/home/standards/certification/iso-survey, accessed 28 June 2016.

and certification procedures would be inconceivable without the comprehensive number crunching that inserts the expert knowledge and private power of management consultants deep into the social fabric. Hence, when the head of an ICT and offshoring industry association claims 'Quality is THE vocation of standards',[13] far-reaching power issues are indeed involved.

Table 7.1 presents the range and core attributes of standards most widely used in the heyday of certification for business processes in Indian customer centres and ITeS-BPO companies. A striking feature is the width of this scope: they not only address quality of management systems and a multitude of tasks performed but also issues related to IT security management and performance targets regarding content. Moreover, it is worth noting that while some of them originate from official standardisation bodies such as the ISO and its alter ego the International Electrotechnical Commission (IEC), many others come from different standardisation bodies and umbrella organisations – something I will discuss at greater length when I examine *who* sets such standards. A major case in point are the tools developed by the Software Engineering Institute (SEI), a research and development centre sponsored by the US Department of Defence and operated by Carnegie Mellon University – now transferred to the CMMI Institute, a technology commercialisation enterprise working as a subsidiary of that private American university. Instead of disciplining management systems, the tools, described as capability maturity models, target the content of business processes in order to assess the ability of an organisation to perform the expected tasks. The fundamental mission of such standards is to provide a tool needed to let service providers and their clients conduct full-scale assessment of the problems likely to occur in all the tasks involved when a business process is performed. This means breaking large and complex business processes into tiny component modules in order to specialise activities. Basically, this implies a description and codification of tasks in their totality, before assessing the extent to which they can be executed using a set of consistent and repeatable steps, i.e. be fully standardised (Paulk, 2001; McIvor, 2010). In practice, this requires detailed documentation of hundreds of pages with countless quantified targets for different issue areas. Developed initially to focus on the managerial dimension of problems encountered by software developments in

[13] Mohammed Lakhfili, Head of Logica North Africa and President of APEBI (Moroccan Federation of Technologies of Information, Telecommunications and Offshoring), interview with the author, Casablanca, 21 October 2009.

Table 7.1 *Quality and security standards most widely used in Indian ITeS-BPO companies*

	ISO 9000 series	ISO/IEC 20000	ISO/IEC 27000 series	CMMI-SVC	PCMM	eSCM-SP	COPC-CSP COPC-CX	Six Sigma	COBIT
Scope	Quality management system	IT-enabled service quality management system	Information security management system	Capability Maturity Model Integration. Service quality management system. Replaces software capability models (SW-CMM)	People Capability Maturity Model. HR quality management system	eSourcing Capability Model for Service Providers	Performance management framework for customer relations services	Quality management system based on defect statistics in IT manufacturing	Control objectives for information and related technology (IT-enabled service quality management system)
Funding	Public + Private	Public + Private	Public + Private	Private + Public	Private + Public	Private	Private	Private	Private
Standardisation body	BSI -> ISO	BSI -> ISO	BSI -> ISO	Software Engineering Institute -> CMMI Institute	Software Engineering Institute -> CMMI Institute	IT Services Qualification Center	COPC Inc. (previously 'Customer Operation Performance Center')	Motorola corporate university	IT governance Institute (ITGI)
Umbrella organisation or company	ISO	ISO/IEC	ISO/IEC	Carnegie Mellon University and US Department of Defense -> ISACA	Carnegie Mellon University and US Department of Defense -> ISACA	Carnegie Mellon University	Private company	Motorola	Information Systems Audit and Control Association (ISACA)

Release date	1994: ISO 9000 2015: ISO 9001:2015	2005: version 1 2011: version 2 2016: under revision	2000: ISO 17799 (replaced in 2007 by ISO 27002) 2005: ISO/IEC 27001:2005 2013: ISO/IEC 27001: 2013	1991: CMM v1.0 (precursor) 2000: v1.0 CMMi 2010: v1.3 for services (CMMI-SVC)	1995: Version 1 2001: Version 2	2001	1996: version 1 2016: version 6 (renamed COPC-CX)	Started in mid-1980s	1996: COBIT 1 2012: C OBIT 5
Geographic Origin	UK -> Global	UK -> Global	UK -> Global	USA	USA	USA	USA	USA	USA

(*Source*: Nasscom Strategic Review 2007, 2010, 2012; interviews and compilation by the author)

defence contracts in the United States, capability maturity models now come in several configurations.

For instance, the latest version of the guidelines for a so-called Capability Maturity Model Integration in the field of services (CMMI for Services) outlines dozens of distinct processes, such as 'Capacity and Availability Management', 'Causal Analysis and Resolution', 'Decision Analysis and Resolution', 'Measurement and Analysis', 'Service System Development', or 'Work Monitoring and Control'. Companies are expected to define relationships between them and find ways to integrate specific practices involved in all those process areas. The ultimate power of the standard is thought to result from the ability of firms to evolve and compete in an ever-changing market thanks to a reflexive behaviour labelled 'continuous and staged representation' (a staged representation is, according to the reference document, concerned with the model as a whole, whereas continuous representation deals with individual processes). In order to measure improvements in processes and allow for competing benchmarking between firms, the model uses so-called capability levels regarding distinct tasks (incomplete; performed; managed; defined) and 'maturity levels' (initial; managed; defined; quantitatively managed; optimising), each of them defined as an 'evolutionary plateau for organizational process improvement' that eventually would characterise the overall performance of the firm (Software Engineering Institute, 2010: 26). Unsurprisingly, with such methodology, the latest version of the guidelines *CMMI for Services* published in 2010 uses more than 500 pages of detailed description.

Besides ISO/IEC standards and capability maturity models for fairly complex business process outsourcing, Table 7.1 shows that service offshoring in India also relies on tools specifically dedicated to more basic call centres and customer relation services, as well as on a flurry of management methodologies and performance tools fiercely competing for the lucrative market of business processes certification. Suffice it here to sketch out the most widely used among them. COPC Inc. (formerly known as Customer Operation Performance Center) sets performance-driven standards specifically devised for customer contact centres. Created in 1996, the first version of its Customer Service Provider (CSP) standard was based on the Malcolm Baldrige National Quality Award criteria and framework; its sixth version was renamed the Customer Experience (CX) standard in 2016, with the intention of giving more emphasis to the idea of a shared partnership – the new buzzword for defining the relationship between service providers and their clients in a supposedly truly co-defined and arguably co-produced service. For its part, COBIT (Control objectives for information and related technology) is a management framework designed by the IT governance Institute, an American non-profit corporation established by a parent professional organisation with a worldwide presence in auditing controls for computer systems. Initially, the tool was principally used by IT auditors, but it has now expanded to include all sorts of metrics related to information security, risk management, and regulatory and compliance issues, such as those required by the Sarbanes-Oxley Act of 2002 that set new and expanded requirements for IT controls and reporting processes. Finally, Six Sigma is a management technique that has been extensively exploited across a wide range of industries. Originally developed by Motorola in the mid-1980s, it is based on defect statistics in IT manufacturing, with the aim of minimising the variability in business processes. To this end, it outlines key performance indicators which, once broken down, can quantify thousands of business processes and reach targets such as reduced costs and/or time, improved customer satisfaction, and, ultimately, increased profit.

Such cases of standards qualifying management systems and the content of business processes indicate how far the material continuum of standardisation brings together the two poles of the physical and societal world in supporting the offshoring of all sorts of IT-enabled services and business process outsourcing. It is true that most of the tools devised for such purpose could be considered to be close to the physical end of the continuum. Indeed, the work processes involved in the definition of the service provision become a simple technical challenge to be solved in such a way as to define segmented tasks on the basis of their lowest common denominator. However, this does not mean that there is no societal dimension involved and no debate concerning their political economy content. First of all, standards do not float in thin air; they are socially embedded and, therefore, when firms adopt them, more often than not they must struggle to get them truly implemented. To this end, mid-level management's challenge is the establishment of a reliable relationship with the workforce. As Arvind Kasi, vice-president for quality & compliance at 74/7 Customer, a customer relations firm, points out, 'the most important thing is that the practical implication of using standards is the need of documentation; standards do not make people straightjacketed and losing opportunities to act … it is not necessary to do all the paper work … and if documentation is a problem, then keep it aside and adopt best practices'.[14] Process standards such as those we have discussed in this section are even identified by some managers of

[14] Interview with the author, Bangalore, 28 January 2016.

the Indian BPO industry as freely available knowledge accumulated over years of innovation in management techniques. At least, this is what Rajesh Segal would have us believe when he considers the leadership taken by Indian companies in instigating a new ISO standard specifically focused on BPO as 'the biggest initiative we could be engaged in ... we wanted to take this knowledge that we have created back to society or to the industry as such.' Moreover, as we will see, with the Indian initiative to launch a new ISO/IEC standard specifically dedicated to IT enabled services and business process outsourcing (ISO/IEC 30105), strong interests oppose two approaches to quality and security standards in the service sector: those in favour of assessing the content of business processes and those behind the certification machinery associated with the ever-growing families of management systems. Last but not least, such quality and security standards, like all standards, draw boundaries between those who conform and those who do not. As Parthasarathy and Srinivasan (2008: 280) remind us, 'since this clearly creates winners and losers, and there is no "best" means of drawing boundaries to benefit everyone equally, standards are socially contested'.

All this provides some evidence that process standards used in India to serve offices around the world extend along a material continuum that ambiguously includes physical and societal dimensions. Where does that leave our second analytical dimension focused on the actors in charge of setting standards?

Who Sets the Standards?

Who plays or claims to play a role as standard setter and thus gains power to define how companies and their employees are expected to conform? In other words, who exactly has the authority to set those standards? Much ambiguity remains regarding the wide range of actors and standardisation bodies which create the tools used to connect the Indian office to the rest of the world. In many respects, the ambiguity blurring the public and the private spheres of the institutional continuum of standardisation enables the authority of standard-setters who would otherwise have more difficulty ensuring wide recognition of the large range of instruments devised for the industry. In the same way as we just saw that standards are seen as a passport to provide 'whatever the client asks for', regardless of their political economy content, standards users in the Indian office of the world give little weight to the institutional nature of bodies in charge of setting standards. It comes as no surprise then that a fragmentation of standard-setters prevails and that the Indian service offshoring industry has for long been confined to the status of standard taker, rather than standard maker. As we will see, the situation might change in the near future as a result of the Indian initiative to develop a new ISO/IEC standard specifically dedicated to IT-enabled services and business process outsourcing (ISO/IEC 30105).

In 2008, in the heyday of the quest of Indian firms for certification of management systems, business processes, and maturity models, mid-level management officials in charge of quality clearly saw no difference of status between ISO standards and management methodologies and performance tools devised by American firms, research centres, or their spin-off technology commercialisation enterprises, such as the CMMI Institute. Thus, Raman Roy, one of the pioneers of the BPO industry in India, could claim: 'Who sets the standards is not important. The most important is what the market needs and the responses given to it.'[15] Similarly, According to Punit Kumar, a general manager for corporate communication at Wipro, 'it may not be possible to have a universal standard in the ITeS industry, because we are all competing for ourselves. ... The customer is the only sacro-saint. If the customer wants us to adopt a standard, we will.'[16] As the 2010s went on, many voices argued that the traditional role of certification to recognised standards may not be as much of a 'differentiator' as it was earlier. Unsurprisingly, the distinct institutional nature of organisations in charge of setting standards was not seen as so important either. In such settings, mid-level management could keep on shopping freely among standard-setters and make the following claim: 'We have all the minimum standards in place ... we can thus provide whatever the client ask for to comply with.'[17] Similarly, when asked about the types of process standards used and their respective significance, in some cases interviewees first flagged ISO tools; in others, business processes and maturity models such as those devised in the United States by the CMMI Institute, COPC, or COBIT were the only ones worth an acknowledgement.

Be that as it may, those different strands of standard-setting bodies are not as opposed as conventional accounts would have it. As we saw in Chapter 4, the ISO and IEC standardisation system follows a model of national participation or delegation, with a national body holding the voting rights used in the technical committees at the international level. In contrast, standardisation in the United States follows a model of direct participation, where companies have direct access to standard-setting

[15] Interview with the author, Gurgaon, 8 February 2008.
[16] Interview with the author, Gurgaon, 8 February 2008.
[17] Manoj Brahmankar, VP for corporate business excellence, HGS (Hinduja Global Solutions), interview with the author, Bangalore, 29 January 2016.

activities fragmented between a number of sectorial organisations competing for market-driven recognition of the international reach of their standards. It is against this background that tools set by research centres and management consultancy firms such as the CMMI Institute can be recognised and valued as highly as standards of the ISO portfolio. Yet, my analysis emphasises that behind labels of 'national delegation' for ISO and 'direct participation' in the United States, actors setting standards are mostly the same; large firms dominate technical committees, with some minor involvement of government agencies and a quasi-total absence of not-for-profit associations from civil society. Standard-setters in the ITeS/BPO industry reflect the same picture.

Although privacy and disclosure rules get in the way of gathering full evidence of industry-level membership and participation, a first approximation of such transnational hybrid authority is the continuing importance of business processes and maturity models devised by US bodies. A striking feature of the CMMI Institute is how it resembles hybrid organisations as described by Koppel (2003).[18] While privately owned (initially by a private university, then its spin-off), it was created at government request to address the specific public policy concern of defects in IT defence contracts in the mid-1980s. It is under such circumstances that Carnegie Mellon University established the Software Engineering Institute in 1984 as a Federally Funded Research and Development Center. Initially, its funding source came mostly, perhaps uniquely, from public funding (the US Department of Defence). As a mix of private initiative and public resources, the entity later expanded by entering the larger market for management standards and business process capability and maturity models; this gave it the opportunity to lessen its dependence on revenues derived from government – a move apparently reinforced in 2016 by the acquisition of the CMMI Institute by ISACA, the professional association for IT governance, assurance, and cybersecurity that also offers COBIT, another widely used standard in the ITeS/BPO industry around the world.[19] Yet, as Koppell (2003: 8) and Weiss (2014: 154) remind us with regard to the difficult task of identifying those hybrid entities operating under some sort of government sponsorship, they essentially rely on the 'functional ambiguity' that allows them to cunningly combine the best worlds of both the private and

[18] Chapter 2.

[19] 'ISACA Acquires Global Capability Maturity Leader CMMI® Institute', ISACA Press Release, 3 March 2016, online: www.isaca.org/About-ISACA/Press-room/News-Releases/2016/Pages/ISACA-Acquires-Global-Capability-Maturity-Leader-CMMI-Institute.aspx, accessed on 28 June 2016.

public sectors. While clearly ambiguous as the public or private spectrum of the institutional continuum of standardisation is concerned, the CMMI Institute is less so, however, in terms of national representation. Although some Indian business pioneers have been associated with the work undertaken by SEI/CMMI, the bulk of the organisation revolves around the large service management firms that belong to the Carnegie Mellon University's network of partners based in the United States.

A second approximation of fragmentation of standard-setters under the ascendency of large American IT service management firms is provided by the lack of involvement of the Bureau of Indian Standards (BIS), the official national standardisation body close to the Indian government. Although claiming membership in hundreds of technical committees and subcommittees, actual participation has so far remained low. The weak involvement of BIS in international standardisation activities is shown by the small number of secretariats of technical committees or subcommittees for which it has responsibility and in which national standardisation bodies are known to have considerable leeway to build an understanding around their schemes. The number of secretariats at technical committee or subcommittee level for which a national standardisation body is given responsibility is often taken as the most appropriate proxy for their influence at ISO (Afnor, 2018). In 2018, BIS was in charge of only ten secretariats, which represents little more than 1 per cent of 800 or so secretariats at work at ISO.[20] In comparison, China, as another large emerging country, was not only involved in sixty secretariats (8 per cent), but took the lead on a number of initiatives directly related to services and management systems (Afnor, 2018). For instance, the China National Institute of Standardisation promoted the revision of the ISO/IEC Guide 76 that provides recommendation on consumer issues to be considered in developing standards for services; it also instigated the creation of a new working group on consumer issues in services (ISO/COPOLCO/WG 18), for which it serves as convener. In this regard, one can understand the aspiration of BIS Director General

[20] In 2018, BIS secretarial responsibilities were in the following ISO committees and subcommittees, some of which bear a striking legacy of the British Empire: ISO/TC34/SC7 – Spices, culinary herbs, and condiments; ISO/TC 113 – Hydrometry; ISO/TC 113/SC 1 – Velocity area methods; ISO/TC 113/SC 6 – Sediment transport ; ISO/TC 120 – Leather; ISO/TC 120/SC 1 – Raw hides and skins, including pickled pelts; ISO/TC 120/SC 2 – Tanned leather; ISO/TC 120/SC 3 – Leather products; ISO/TC 146/SC 1 – Stationary source emissions; ISO/IEC JTC 1/SC 7 – Software and systems engineering (see: https://www.iso.org/member/1794.html?view=participation&t=S, accessed on 9 August 2018).

Alka Panda 'to become more pro-active in ISO by participating to more TCs.[21] According to J. Roy Chowdhury, BIS Head of international relations and technical information services, the challenge that lies ahead for BIS is indeed clearly to 'play the role India is expected to play as a large emerging country but that it is not playing now'.[22]

In short, by examining who exactly has the authority to set standards most widely used in India to service the offices of the world, we see much ambiguity in the fragmentation of the bodies involved and actual actors drafting the specifications. While the industry hardly makes any distinction between ISO/IEC standards and the management tools and business process methodologies devised by American private or hybrid organisations, the prominence of large firms and consulting companies and the weak involvement of the Bureau of Indian Standards indisputably tilts the balance of the institutional continuum of standardisation towards the private sector.

Where is the Indian Office of the World Standardised From?

The third axis of our analytical framework is the spatial continuum where the jurisdictions that support the system of recognition of standards overlap. Earlier chapters have shown that, here too, standards are ambiguous: they rest on the dual nature of sovereignty – the principle of the territorial state on which lies the endogenous recognition of standards and the exogenous processes of the transnational guarantees given to the principle of contract inviolability in a world of globalised capitalism. We have just seen that attempts to homogenise management tools and business process capability and maturity models across sovereign spaces face a plurality of standards. The question I now address is how do the intertwined exogenous and endogenous poles of the spatial continuum of standardisation play out in the ability of the Indian ITeS/BPO industry to use and comply with standards.

A first point is worth mentioning: the fact that Indian ITeS/BPO industries are identified as standard takers, rather than standard makers undoubtedly encapsulates a spatial dimension. At the time when management systems and business process standards became all-powerful, several Indian business executives and standardisation officials were already concerned by the dominance of imported standards initially

[21] Alka Panda, BIS Director General, interview with the author, New Delhi, 21 January 2016.

[22] J. Roy Chowdhury, BIS Head of international relations and technical information services, interview with the author, New Delhi, 21 January 2016.

conceived for IT manufacturing and basic service outsourcing within the United States. The absence of industry-wide standards specifically dedicated to the ITeS/BPO sector was seen as a typical case in this regard. Whilst the situation obviously resulted from the American first-mover statute in outsourcing services, Indian entrepreneurs and officials were worried that it could generate difficulties for further consolidation of the sector in India. According to Rama Mohan, Head of Business Transformation Group at Infosys BPO, 'for the whole BPO industry ... all standards adopted are global standards, with no Indian origins ... there is a need for the Indian model to become a new global model, in which the Indian perspective could be brought.[23] High-ranking officials at the Bureau of Indian Standards clearly shared this view when they claimed to be 'bothered that standards are imposed [and suggested that] BPO industries in India should become standards makers and make the standards themselves instead of taking them'.[24]

Arguably, no tool other than the so-called eSCM standard provides evidence of such exposure to the exogenous logic of market recognition and dependency in Indian service offshoring. eSCM stands for eSourcing Capability Model. It is a framework developed in the United States by ITSqc – another spin-off from Carnegie Mellon University. Developed as a model specifically dedicated to IT-enabled services industries to improve their relationship with their clients, it was initially seen as a strategic tool for Indian firms that would help them to scale up the value chain and keep competing at a global level. As Manoj Brahmankar, Vice President for corporate business excellence at HGS (Hinduja Global Solutions), points out, 'we adopted eSCM ... we found a lot of value in terms of practices across the life-cycle'.[25] Yet, the same informant somehow reluctantly later made us understand that, 'unfortunately, the standard is not being maintained anymore; any standard that is not maintained in the business environment does not stay relevant'. Clearly, this means that cost and energy spent in reengineering the organisation of the company so as to comply with the standard was basically undertaken in vain – a situation that many companies in India and elsewhere have experienced as the tool became obsolete. According to Rajesh Segal, Head of Quality and Process Excellence at Wipro, the reason why 'eSCM didn't take off that well [is that it was] too detailed

[23] Interview with the author, Bangalore, 11 February 2008.
[24] Rakesh Verma, additional Director General of the Bureau of Indian Standards, interview with the author, New Delhi, 6 February 2008.
[25] Interview with the author, Bangalore, 29 January 2016.

and heavy on practices'.[26] Similarly, the former Nasscom vice president for BPO describes eSCM as 'very bulky to implement; you would need an army of people to keep track of the data and information in order to get the certification'.[27] What was clearly a painful experience with wide consequences for any organisation having adopted eSCM also provides harsh evidence of how firms, as standards takers rather than standards makers, adopt and comply with standards set and maintained in organisations far abroad. As a result, they depend on exogenous developments and have no leverage whatsoever on their outcome.

Whilst the exogenous logic of transnational market recognition backed by the dominance of US-imported IT management and business process standards for the most part prevails, it is important to note the endogenous dimension upon which such developments relied at first. Building on the approach sought after by the aforementioned 1994 World Bank study calling for a transition towards the 'low cost, high quality' quadrant of the global market, the first wave of certification of the Indian industry targeted the newly established ISO 9000 total quality system management standard. And here it was the territorial state that was called in to support the required market recognition. The Government of India launched a programme of subsidies, with firms awarded ISO 9000 or equivalent certification made eligible for a grant from the state-owned Export-Import Bank, thanks to which they could claim up to 50 per cent of the costs of obtaining quality certification (Sharma, 2015: 176–177). Such a finely tuned policy supporting the early development of Indian and foreign affiliate ITeS/BPO firms shows that the shift towards liberalisation policies that took place after 1991, far from leading to a diminishing role of the state, rested on a number of targeted interventions among which certification subsidies, skill development, marketing assistance, and training were particularly instrumental for supporting the authority of standards.

From Standard Takers to Standard Makers: The Power of Nasscom

As stated earlier, Nasscom is the National Association of Software and Services Companies – the Indian industry body that has supported the Indian ITeS/BPO companies to compete on the global market and set

[26] Interview with the author, Sarita Vihar (Delhi), 22 January 2016.

[27] Raju Bhatnagar, Secretary General of the Bangalore Chamber of Industry and Commerce and former Nasscom vice president for BPO, interview with the author, Bangalore, 27 January 2016.

very high standards for themselves. In many respects, Nasscom alone epitomises the ambiguous power configuration supporting the recognition of standards. Established in the late 1980s by a small group of Indian entrepreneurs active in the United States as a way to overcome the mutual mistrust that prevailed between government and the IT and software industry in India, it became so successful that it soon substituted for government policies in many issues closely or loosely related to the industry. Basically, the government has given Nasscom a blank cheque to develop industrial policy at home and promote the industry abroad. While formally a private not-for-profit business association, it executes all sorts of policies that would otherwise be under the sole responsibility of the state. Nasscom reflects intrinsic ambiguity regarding its position in relation to the private and public spheres. Moreover, while the body was first instrumental in supporting narrowly defined system management standards, it has later widened its activities to broader societal issues, including security concerns and privacy protection. Finally, it demonstrates considerable ambiguity by mingling the spatial dimensions on which the recognition of standards rests. As we have just seen, as an Indian business association, it has a close relationship with the territorial state that reinforces the endogenous principle of standards recognition; at the same time, an essential part of its work is to support the exogenous practices through which US-imported IT management and business process standards used in India may gain transnational market recognition.

In the 1990s, besides lobbying the government for major reforms in legislation, it joined forces with the government's incentive programme to make compliance to quality standards a top priority. Throughout the 2000s and 2010s, it continued to play a major role in institutional reforms and standards promotion. For instance, discussing the major step characterised by the adoption of the IT Act in 2000 regarding the inclusion of new security guarantees, a high-ranking official who lived through the early history of Indian IT bluntly claims: 'basically, Nasscom drafted the Act'; and when the 2008 Data Security Act came as a new piece of legislation intended to fill the gaps left by the previous one, our informant stresses that 'here again Nasscom had a major drafting role'.[28] Just as Nasscom has been the main driving force behind the most important regulatory oversight initiatives of the industry, it also took operational responsibility in setting up and collecting data privacy and security standards. Of note in this regard is the creation in the second half

[28] Puneet Kumar, General Manager for corporate affairs, Wipro, interview with the author, Gurgaon, 8 February 2008.

of the 2000s of the National Skill Registry despite a number of heated discussions regarding implications for recognised standards of privacy protection. As service offshoring in India deals with sensitive data not only from client companies but also individuals, for instance with health-care records and tax forms, Nasscom collects all sorts of information, which allows employers to perform background checks on existing or prospective employees. At about the same time, it also set up the Data Security Council of India (DSCI), a self-regulatory agency to uphold data privacy and security standards.

Many accounts of the Indian outsourcing success story put Nasscom in the limelight. In the words of Kshetri and Dholatia (2009: 231), much of the credit for the remarkable progress in Indian offshoring firms' success should indeed go to non-state actors such as Nasscom, whose role is viewed as 'phenomenal … in monitoring the industry behavior and bringing significant institutional changes'. The standing of Nasscom as the successful voice of India in support of the ITeS/BPO industry remains, however, in debate. First of all, local interests are often put on the back burner. The ability of Nasscom to work closely with central government and provincial state officials has in numerous cases led to stiff opposition from important parts of the population that depend on the traditional economy and on access to land pre-empted by Software Technology Parks and other comparable developments. While the land issue is one of the major bones of contention, the rapid rise of such enclave economies within a predominately agricultural society with a long history of social justice movements and conflicts on caste identities has led to a number of wider dislocations, be it in Bangalore, seen as the Mecca of global service offshoring, or in many other locations elsewhere in India (Upadhya, 2009). Important, too, is the fact that trade unions are systematically excluded from industrial relations in service offshoring. The IT, software, and ITeS/BPO industry does not report under the Companies Act that frames the business environment in India, but under the Shop & Establishment Act, which provides large exemptions regarding industrial relations issues. In addition, in most states across India and for most of its existence, the industry enjoyed exemption from the labour law.[29]

Second, while conventional views claim that Nasscom played a crucial role in building the trustworthiness of Indian offshore services, critiques point out that the support given to quality standards and initiatives such

[29] Industrial Employment (Standing Orders) Act, 1946; see, for instance: 'Labour unions cry foul over exemption of IT from labour law in Karnataka', *The Economic Times*, 7 November 2013.

as the National Skill Registry epitomises the cartelisation of the industry, levelling the employment playing field and agreeing not to poach their respective workforce between themselves. Perhaps more importantly, it reflects its anti-union stand. From this point of view, the registry is rather a blacklist of employees whose potential implications with unions and industrial action are labelled as managerial risk. This is obviously the backdrop that leads Karthik Shekar, General Secretary of the National Confederation of Unites (NCU), to underscore that 'Nasscom's registry is used to create fear among workers, a psychosis among those who know that they are blacklisted as not a single company will accept their application whatever their skills are.'[30] This pioneer union activist in service offshoring in the Bangalore region shows great concern that despite complaints made up to the Prime Minister's office, no legal action can be taken, as cases are impossible to document in accordance with courts' requirements – a situation that helps Nasscom to continue claiming that such allegations are groundless. It is no wonder that the activist does not have a high opinion of quality standards either: 'managers bring in fancy jargon like CMMi and the like, the middle management, with western culture education, but deeply Indian, ends up totally confused. Call it pcmm-3, but it's just a matter of how people work and the reality on the ground is completely different.'[31]

A last point is the difficulty of distinguishing between the domestic and foreign interests represented in Nasscom. The body persistently defines itself as truly representative of the industry. A high-ranking official can thus proudly claim a membership of '97% of the IT industry … consisting of Indian companies but also foreign affiliates [standing united behind] a body defining and promoting self-regulation'.[32] Conventional analyses are keen on taking over this assumption. For instance, Kapur praises the role of Nasscom that, unlike other industry associations in India and many other developing countries, is neither distant from the state nor in continuous conflict with its members, thus 'giving the industry a unified voice [working] in tandem with the Indian state to jointly promote the sector's interests' (Kapur, 2002: 98). In contrast to this view of a body working hand-in-hand with domestic firms and affiliates of foreign companies to support the sector's interest, some contrasting voices assume that Nasscom is – or at least was at the beginning of its existence – rather a mole working in the interest of large American IT

[30] Interview with the author, Bangalore, 26 January 2016.
[31] Interview with the author, Bangalore, 13 February 2008.
[32] Ameet Nivsarkar, Nasscom Vice President and Head of Research, interview with the author, New Delhi, 7 February 2008.

and service firms. According to Rajesh Kalra, Chief Editor at the The Times of India Group, there are indeed many allegations about Nasscom: 'At the beginning, it was controlled by a few US companies who put a lot of money in it and Nasscom eventually helped them a lot … basically, it's a big boys' club of the rich and powerful … a lobby for foreign captive companies … Google, Microsoft and so on have strong leverage in India'.[33]

After all, whatever the endogenous or exogenous dimension, the technical or societal nature, and the private or public purpose of the interests pursued by Nasscom, the umbrella association clearly understands the importance of standards for the ITeS/BPO industry in India and the risks associated with a fragmentation of instruments with more or less overlapping, duplication, and dependency. As of 2007, a group of far-sighted Indian entrepreneurs perceived the need for a standard that would be specifically dedicated to the BPO industry and was likely to be adopted independently of decisions taken thousands of miles away in any subsidiary or spin-off of a private American university. As the respected pioneer of the industry Raman Roy pointed out when I met him in early 2008, India must 'take the lead; Carnegie Mellon University set business processes standards for IT, Nasscom should do the same for BPO, with, for instance, Nasscom standards level xyz. That kind of standardisation is now critical for our growth rate targets.'[34] Nasscom indeed took over the initiative to put India in the driving seat and transform the industry into a standard maker rather than just a standard taker. This is basically how the ISO/IEC 30105 standard was born and eventually published in 2016 under the general title 'Information technology — IT Enabled Services/Business Process Outsourcing (ITESBPO) Lifecycle Processes'.[35]

How ISO/IEC 30105 Came to Life

Fixing the Y2K bug at the turn of the millennium provided Indian IT services with an opportunity to prove itself to its clients with all the required certifications in place. In the following years, the BPO industry in India had already matured and was widely seen as a credible destination. Although a number of standards were used, such as eSCM and COPC, none of them were as widely accepted as CMMI tools used in the

[33] Interview with the author, Noida (Delhi), 19 January 2016.
[34] Interview with the author, Gurgaon, 8 February 2008.
[35] Initially within the subcommittee on software and systems engineering (ISO/IEC JTC1/SC7/WG25) and subsequently transferred to the subcommittee on IT Service Management and IT Governance (ISO/IEC JTC1/SC40/WG3).

software industry. As emphasised by Raju Bhathnagar, first convener-to-be of the ISO/IEC working group, 'there was a gap which needed to be filled – that is a standard for the BPO industry itself'.[36]

Nasscom took advantage of a plenary meeting of an IEC/ISO technical committee organised in Hyderabad in May 2008 to submit its proposal for a new item agenda. Badly prepared, not attuned to ISO/IEC procedures, the move was sent back to the drawing board. A study group eventually convened to better determine such needs. With the help of a few experts sharing their previous experience with foreign national standardisation bodies, Nasscom steamed ahead to achieve its plans. It forcefully lobbied the Bureau of Indian Standards (BIS), who only had a limited understanding of the project. It thus made sure that BIS, as the national standardisation body representing India in IEC/ISO arenas, would be in a position to take the lead on the project. In order to make its case for an IEC/ISO standard dedicated to BPO only, it enrolled close to a hundred stakeholders at home and abroad among affiliates of foreign companies and national standardisation bodies. In 2009, when it came back to the negotiation table, it was much better prepared and could provide sufficient evidence of a business case for a new formal standard.

While eventually adopted, the proposal to establish a new standard (ISO/IEC 30105) still faced considerable resistance. The group of experts that a few years earlier had launched the first version of ISO/IEC 20000 on IT service management were at odds with what they identified as a proposal containing too many overlaps with their own project so as not to compete with it. According to one of the top experts in charge of the working group responsible for ISO/IEC 20000, a key concern was accordingly to make sure to be in a position to control any future developments[37]. ISO/IEC 20000 thus undertook some manoeuvring to change the organisational structure in which the drafting of standards would take place by creating a new subcommittee to accommodate both working groups responsible for ISO/IEC 20000 and ISO/IEC 30105-to-be. Such a turf war concealed a deeper conflict between two opposing approaches to quality and security standards: on the one hand, those in favour of keeping the definition and control of the tasks at the outer layer of management systems, and those who advocated digging into the business processes models themselves on the other. According to an expert who was present throughout the deliberations, the Indian initiative prompted such an 'interesting, or even a virulent debate' precisely because some delegations, in particular from France

[36] Interview with the author, Bangalore, 27 January 2016.
[37] Confidential interview with the author, Geneva, 13 June 2016.

and, to a lesser extent, the United Kingdom, were strongly opposed to the Indian approach based on business processes, seen as incompatible with the basic principles guiding the certification of management systems.[38] In contrast, the Indian view was that 'in BPO, process implementation and value creation (i.e. internal process bringing value and ability to improve it) are key'.[39] It is difficult to come to any final analysis on whether the debate was guided by personal agendas, certifiers' interests, matters of principle, or simple misunderstandings. However, when the time for a decisive vote came, it was only by the abstention of one expert who saw no contradiction between both methods that the Indian approach was eventually approved and the project launched.

After more than six years of drafting procedures, ISO/IEC 30105 was published in 2016 under the name 'Information technology — IT Enabled Services/Business Process Outsourcing (ITESBPO) Lifecycle Processes'. It consists of five parts with the aim of serving as reference, assessment, and maturity models and guidelines for service providers performing outsourced IT-enabled business processes. Figure 7.1 outlines the relationship between the various dimensions included in the main parts of the standard. Delving into the technical detail of the document would take us beyond the scope of this book. Suffice it here to stress that a striking feature of the standard is at once its distinctiveness and its comprehensiveness. Although initially based on the ill-fated eSCM standard of the CMMI family with a similar two-dimensional model of process categories and organisational capability levels, the four parts of the standard describe a set of detailed tasks, processes, evaluation procedures, and organisational maturity. Ideally, these cover the whole lifecycle of services provided by business process outsourcing companies in areas as diverse as human resource management, administration, health care, banking and financial services, supply chain management, travel and hospitality, media, market research, analytics, telecommunication, engineering, and manufacturing.

The story of the ISO/IEC 30105 standard detailed here shows that Nasscom put considerable energy into shepherding the project of a new ISO/IEC standard specifically dedicated to the BPO industry through to a successful conclusion. If nothing else, it provides compelling evidence of how standards are identified as powerful tools in the organisation of

[38] Alain Renault, Senior R&D Engineer at the Luxembourg Institute of Science and Technology, former member of ISO/IEC JTC 1/SC 40 WG 3, IT-enabled services/Business process outsourcing, Skype interview, 16 June 2016.

[39] Ravi Veeraraghavan, Vice President for Business Process Outsourcing, TCS; convener of ISO/IEC JTC 1/SC 40 WG 3, IT-enabled services/Business process outsourcing, Skype interview with the author, 21 June 2016.

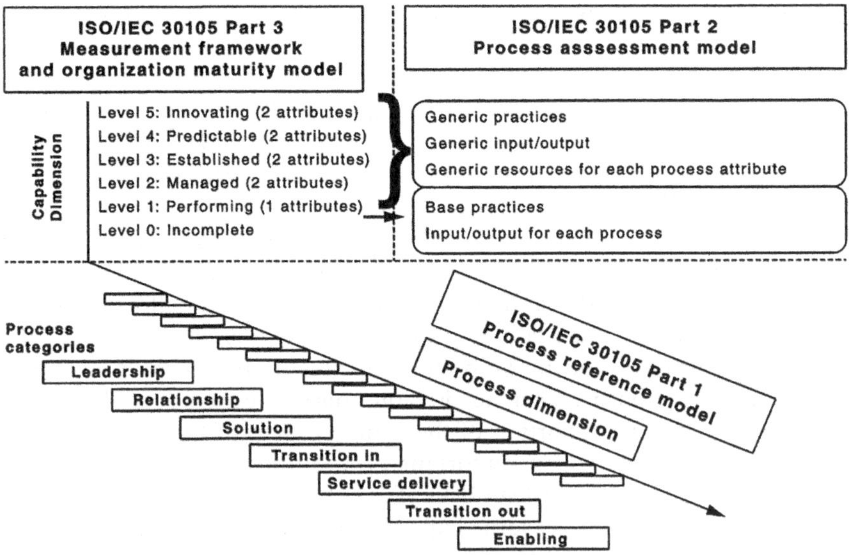

Figure 7.1 ISO/IEC 30105 and its various parts
Source: Sample Draft International Standard ISO/IEC 30105, ISO/IEC JTC 1/ SC 40, reference number ISO/IEC DIS 30105–2:2015(E), 2015.

global markets and for which valuable resources are devoted to move technical diplomacy into full-swing. The choice of targeting the joint ISO/IEC arena demonstrates that private-only instruments *à la* CMMI have their own limits. The hybrid public–private dimension of ISO/IEC arenas is seen of higher standing and worth spending substantial resources to draft instruments that can be explicitly recognised as international standards in the sense of Article VI: 5b of the General Agreement on Trade in Services (GATS). Moreover, ISO/IEC 30105 describes highly technical specifications – a scientific organisation of business practices that remains, however, deeply embedded in societal values. As a method devised to 'help companies achieve transformational outcomes more quickly',[40] the detailed tasks, processes, evaluation procedures, and organisational maturity levels defined by the standard have extensive impact on the social organisation of the work undertaken within industries and beyond on the everyday life of employees themselves. Finally, by turning the Indian service offshoring industry into standard maker rather than standard taker, ISO/IEC 30105

[40] ITES/BPO Study Group Presentation to WG25, Niigata, 25 May 2010 (PowerPoint presentation).

contributes to shift the intertwined poles of the spatial organisation of standardisation towards its endogenous end. It clearly results from a strong Indian will to gain more recognition of its own ability to innovate in the establishment of a new standard specifically dedicated to the BPO sector of the industry; as one informant put it, a key objective was indeed to 'provide legitimacy to the internationalisation of the Indian outsourcing industry'[41] – as if this marked a new step in the process begun in the 1990s, when the industry had to prove its conformity to recognised standards in order to make service offshoring a legitimate and business-as-usual strategy. At the same time, however, the whole effort would be in vain if, once the standard is published, only a small fraction of the industry in India adopts it. To be fully successful, the move from standard taker to standard maker needs large market recognition of the new standard far beyond a portion of the Indian offices of the world. This is why Nasscom planned an important promotion campaign in 2016 to support swift adoption of ISO/IEC 30105.[42] The campaign was clearly all the more important, as some experts who supported the project in its early days later admitted to having no expectation whatsoever of large adoption of the standard by a service industry already overloaded by the many new regulatory requirements associated with the post-crisis era.

⋆ ⋆ ⋆

The story of how India has become the world office outlined in this chapter shows that standards mattered right from the start of the journey to the latest prospects of the industry. This clearly contradicts conventional accounts that consider the standardisation of tasks in ideal-typical service industries unlikely. Although varying in degrees, business services offshored in India are indeed much more intangible and relational than (re)insurance studied in previous chapters. They are also more labour intensive. And Yet! They seem rather less resistant to standardisation, trade, and internationalisation. The many ways in which standards play a role in fixing quality and security uncertainties thus support my extensive hypothesis on the power of standards in the global expansion of services.

While this account deviates from studies assuming that industry specificity is likely to dictate its propensity to be standardised and, hence, internationalised, what lessons can be drawn from this argument

[41] Alain Renault, Senior R&D Engineer at the Luxembourg Institute of Science and Technology, Skype interview with the author, 16 June 2016.

[42] Ravi Veeraraghavan, Vice President for Business Process Outsourcing, TCS; convener of ISO/IEC JTC 1/SC 40 WG 3, IT-enabled services/Business process outsourcing, Skype interview with the author, 21 June 2016.

regarding the ambiguity supporting the transnational hybrid authority of such standards? At first sight, the power of standards in use in the office of the world looks rather unambiguous: large American IT services management firms act as *de facto* private standard setters for a flurry of instruments focused on highly technical and narrowly defined management systems and business processes; this would leave the service offshoring industry in India as a mere standard taker, dependent on global market recognition whatever its immensely successful accomplishments in gaining certifications to those standards. My analysis has shed light, however, on a different picture in answering the three question of who sets what standards from where. From this perspective, business management and process standards reflect a more ambiguous form of authority that is not thoroughly private, let alone exclusively public, that includes major socio-political concerns behind their thick veil of technical specifications, and whose exposure to an exogenous logic of market recognition intermingles with the endogenous dimension of government subsidies and incentives. The chapter provides ample evidence of how the tools used for the offshoring of business services in India span those three dimensions I refer to as institutional, material, and spatial continuums. Depending on the distinct standards concerned, location on the continuum indubitably varies. But of late, the consolidation of their transnational hybrid authority seems to follow a coherent path towards a sturdier combination of both poles of each continuum.

Regarding the issues included in the material continuum of standardisation, the new IEC/ISO 30105 standard explicitly seeks to broaden the scope of its tool with a target of content defined in the most comprehensive way through the whole lifecycle of business processes. More alarming are the contentions about the anti-competitive, anti-union stance of Nasscom's policy and instruments and its disregard for local people and interests. Moreover, it is worth reiterating that complying with standards draws at the same time hierarchical boundaries excluding those unable to comply. In any case, what is important for our purposes is that fundamental human and labour rights are dealt with in the name of technical standards on business process quality, capability and maturity models, data security, and privacy protection. The ambiguity of issues at stake supports Nasscom's ambitions to use the power of standards to make the Indian service offshoring industry a world powerhouse. Concerning the institutional continuum of supporting the technical diplomacy of standardisation, the resources that Nasscom devoted to support the adoption of a new standard within the IEC/ISO arena in itself shows limits to the private authority of those American entities devising the management methodologies and performance tools most widely used in the industry. Moreover, we saw that those entities cannot even be defined as private, since their functional ambiguity reflects a hybrid pattern able to combine the best of both the private and public worlds. Finally, regarding the spatial continuum along which compliance to standards is recognised, the weight of transnational capitalism obviously keeps tilting the balance towards the exogenous end of the continuum. The prevailing fragmentation of certifications that provide 'whatever the client asks for' pays scant attention to the institutional nature and political economy content of standards. Besides a handful of government policies supporting the adoption of standards and the recent IEC/ISO 30105 initiative expected to demonstrate the ability of Indian service industries to be standard makers rather than standard takers only, the endogenous dimension of standards recognition remains weak. Just as any other form of private authority in the global political economy, if the compliance to standards is only guided by the exogenous principle of transnational capitalism, with inadequate state support and consent, the odds are that it may prove to be weaker than expected. A first step to cope with such a situation would obviously be to rest the authority of standards on a wider segment of the continuum with stronger government policies, a greater involvement of the Bureau of Indian Standards (BIS) in international arenas, and an effective participation mechanism for neglected local people and interests, supporting the ability of civil society associations to bring forward standardisation issues in relation to the everyday life of millions of people.

8 Conclusions

Following the meeting of the Group of Twenty (G20) in Hamburg, Germany, on 7–8 July, 2017, the American National Standards Institute (ANSI) issued a press release to support the final declaration, of which paragraph 11 'encourage[s] the development and use of market- and industry-led international standards for digitised production, products and services' (American National Standards Institute, 2017a). ANSI was right to be pleased! The careful wording used in the declaration not only matches its view on standardisation; with the G20 summits having reached centre stage on international economic governance in the wake of the global economic crisis, it is likely to become a significant landmark for future policy orientations regarding service industries increasingly embedded in IT-related technologies. This is just one more example of the non-conventional forms of power that international standards reflect in the organisation of contemporary capitalism. In many ways, the enquiry which underpins this book has carried me on a long journey with multiple detours and uncharted waters. They include the wide range of low- and high-skill activities provided by the Indian 'office of the world', the new frontiers of insurance markets being found with ageing populations and increasing natural catastrophes, the particulars of prudential standards devised for doubling the security of insurance policies, the intricacies of the European and American standardisation systems, and the likely impact of the new generation of preferential trade agreements on standards. To understand the relationship among globalisation, the expansion of services, and the power of standards, this book proposes several responses to the three questions that have guided my analysis from the beginning: what non-conventional form of power do international standards epitomise in the organisation of contemporary capitalism? Why have they become such prominent tools in global governance? Could they become as prominent for the service sector as for manufactured goods? Before tackling larger implications of my analysis, let me start by recalling my main arguments and the evidence provided throughout the book.

The book has proposed three arguments inferred from the power of ambiguity, the ambiguity of standards, and the rise of services. The first argument is that the pervasive influence of standards rests on the ambiguous dimension of non-conventional forms of power. While a number of scholars in international relations and social sciences use the concept of hybrid to characterise such transformations of power in world politics and patterns of regulation in contemporary capitalism, such use usually remains on a second-best – or default – basis, by emphasising the rise of private actors and standards, as well as the elusive combination of contradictory logic at work in fluid configurations of power. In such a 'neither/nor' context, the idea of hybrid may support ill-defined global governance policies that enable the exercise of authority in wide-flung areas without full attribution of sovereignty. At the same time, it leaves the ontological properties of such non-conventional forms of power and regulation virtually undefined. By drawing on insights from critical approaches in international political economy, semiotics, studies in science, technology, and society, and post-colonial approaches, the book shows that the semantic field of hybridity actually conveys an overall substantive attribute, ambiguity. The concept of *transnational hybrid authority* describes such an ambiguous juxtaposition of instances of power that confers authority on new actors and new issues across sovereign spaces even as they transform the relationship between transnational capitalism and territorial sovereignty. In contrast to most studies on transnational private governance, my analysis is not only concerned with the private or public status of actors and organisations likely to exercise a recognised authority in international affairs. By bringing together the new range of state and non-state actors involved, the scope of the issues on which they operate and the spaces through which their authority may be recognised, my three-dimensional analytical framework includes all three categories of the subjects, objects, and spaces of authority. This approach charts a comprehensive analysis of global governance and the broad spectrum of power instruments it conveys, such as international standards.

So why have standards become so prominent? The book provides ample and diverse evidence of how ambiguity may lend support to the transnational hybrid authority of standards. In contrast to neo-institutionalist analyses focused on how supply and demand factors should meet to make standards alternative forms of private and voluntary regulation, my account demonstrates the ambiguous content of power relations in the regulatory authority of standards across borders, their need in creating new markets, and how this may yield substantial struggles to define and conform to them. It can be characterised as a social institution in its broadest sense only because it introduces a distinct form of domination in the organisation of contemporary capitalism. Following my three-dimensional analytical framework, such ambiguity helps overcome the distinction between the public and private spheres in which mandatory regulation and voluntary standards are usually confined. There is arguably no better case than the European system to epitomise such endeavour to blur the public and private poles of the institutional continuum of standardisation. Ambiguity also supports the surge of all sorts of new and increasingly ubiquitous standards along a material continuum. Like Latour's quasi-objects, all standards link technical and physical specifications to contentious social values and institutions; but some more – and more explicitly – than others. We saw for instance that data exchange standards used in the insurance industry do not just solve interoperability issues; they tackle the content of risks covered too. Finally, we saw that compliance to standards spans a spatial continuum across multiple jurisdictions. Here again ambiguity helps conflate two interlocking principles of the dual nature of sovereignty: the endogenous logic of territorial sovereignty, on the one hand, and, on the other, the exogenous logic reinforcing the transnational underpinning of contract law for capitalist markets. Besides the extent to which ISO and European standardisation systems differ from the American one – let alone the jungle of management tools and certifications – this is why, to take just one example, the mechanisms of regulatory cooperation designed in the new generation of preferential trade agreements raise so much concern. There is a good chance that bodies such as the Regulatory Cooperation Forum established by the Canada–European Union Comprehensive Economic and Trade Agreement (CETA) will trigger concealed transfers of authority outside existing state regulatory agencies and recognised standards-developing organisations.

Finally, could international standards be as important for the service sector as for manufactured goods? A conventional answer would be: probably not – pointing to the intrinsic characteristics of many services, asymmetries of information, uncertainty, or institutional embeddedness for market regulatory or public interest purposes. Such restrictive hypotheses view the standards-defying nature of services as an obstacle to their internationalisation, as shown by the roughly unchanged 25/75 ratio of trade and foreign investment to overall employment and GDP figures in services over the last quarter century. We saw, however, that during the same period developing and emerging countries have practically doubled their share in trade and investment in services. Furthermore, such activities are increasingly diversified, sophisticated, and deeply integrated with manufacturing processes related to global production networks. It is not just the economy that is more services-dependent on an increasingly global level; services themselves are standards-dependent. They rely on standards to respond to all sorts of quality and security requirements.

My argument on the rise of services is thus underpinned by an extensive hypothesis, according to which the ability to set standards for services depends less on distinct sectorial characteristics and domestic institutions than on the power of hybrids, i.e. the ambiguous status of actors setting such standards, of the issues eventually standardised, and of the space on which they are recognised. From this perspective, service standards per se neither hinder nor support a global integration of services; they rather prompt contradictory forces and opposing political economy objectives. Unsurprisingly, their content has never been and will never be a matter of science-based consensus, as shown by the long-standing divide between imperial and metric units. A broader case in point has been the struggle between advocates of so-called horizontal and vertical standards in setting the agenda of the European standardisation strategy regarding services. While the former, chiefly represented by the powerful British Standards Institution (BSI), are in favour of only defining generic attributes supporting a market-based management standard system available on a horizontal basis across the widest range of services, the latter (supported by a majority of European standardisation bodies) give preference to more substantial technical specifications likely to be applied on a distinct sectorial basis in order to define how services can be co-produced and used reliably with shared expectations regarding their quality. We saw that in the end the question boils down to what a service is and is meant to be, and that, considering the importance of the issue, the European Commission could only suggest a 'hybrid combination' in the hope of overcoming such antagonism.

Looking more closely at the assumption that the ability to define standards supporting the internationalisation of services does not vary with the specificity of the activity or domestic institutions concerned, the book studies two contrasted cases, namely insurance and business process outsourcing. Both can be viewed as 'crucial cases' (Eckstein, 1975), in the sense that they score either low or high values on the main characteristics differentiating the service economy, and so can represent the most- and least-likely cases of service standardisation, respectively. Without here repeating the detail and nuances of both cases, my account has provided evidence in support of my extensive hypothesis. While the study of the insurance industry concerns activities far from the ideal type of relational, non-material services, we saw that setting standards for such a most-likely case remains fraught with difficulties, notwithstanding their significance for market creation and regulation. Conversely, the wide range of standards existing in the ideal-typical case of intangible and relational business process outsourcing is also counter-intuitive, as it concerns a least-likely case and thus supports my argument that the power of standards in the globalisation of services should not be viewed as sector specific. Moreover, my analysis, essentially based on the examination of how India has become a world office, has shown that the power of those standards is more ambiguous than is usually assumed in terms of public involvement, societal implications, and territorial recognition.

What, then, are the broader implications of conceiving the power of service standards as a transnational hybrid authority defined by its constitutive ambiguity rather than by sectorial or institutional specificities? I will return to the issues of power and ambiguity, of the ambiguity of standards, and of the rise of services, with distinct reference to the contrasted industries chosen for each case studied.

First, regarding the relation between power and ambiguity, the emphasis on ambiguity might arguably be considered a truism from a linguistic or semiotic perspective. Language and signs do not convey transcendent meanings. Their ambivalence is part and parcel of any semantic field whose plurality of simultaneous readings are, as seen in Chapter 2, what Bakhtin (Bakhtin, 1981) referred to as hybrids. There is more to it, however. Ambiguity also entails a pragmatic dimension. Just as Bakhtin saw the reconstruction of language by novelist as an intentional hybrid serving his creative strength, the power of ambiguity opens up several alternatives in practice and expands the range of actors likely to act. Needless to say, this is widely used in politics and in diplomacy. Days after the unexpected election of Emanuel Macron as French President in May 2017, *L'Obs* (a weekly with a large readership in France) quoted the well-known aphorism of Cardinal de Retz (1613–1679) 'on ne sort de l'ambiguïté qu'à ses dépens' in the irony made about a conversation picked up between the new president and his minister of foreign affairs on a first meeting with trade union leaders, in which he said: 'It went well … Er! … I told them nothing.'[1] As Villar (2005: 60) emphasises, ambiguity 'appears functional because it creates space to manoeuvre'. Jegen and Mérand (2014) show for instance that ambiguity lies at the core of the communicative strategy used by political entrepreneurs in the construction of coalitions in European public policy. From a broader anthropological perspective, Mallard (2014) thoroughly studied the role that ambiguity – in contrast to transparency and opacity – played in the counter-intuitive

[1] *L'Obs*, No 2743, 1 June 2017, p. 14; the aphorism could be translated as 'You cannot get away from ambiguity without damage.'

vagueness of international legal commitments surrounding the Treaty on the Non-proliferation of Nuclear Weapons. Another case in point is the purposeful governance strategy of financial institutions that accommodates distinct forms of ambiguity so as to increase tolerance towards a diversity of expectations (Best, 2005: 28–32).

Besides routine political and diplomatic practices, such ambiguity should not be considered only as a challenge to sovereignty, democratic representation, and the interstate system. As briefly mentioned at the end of chapter 2, it has broader implications for critical approaches in the field of international relations. It should help clarify the 'range of possible alternatives' (Cox, 1981: 130) away from the prevailing order and, therefore, facilitate an assessment of the current potential for emancipatory transformation and change. The holistic historicist epistemology of critical approaches opens up spaces of progressive practices in the face of structural forces and temporalities of *longue durée* not even thought of otherwise. This work has been very important in its own right. Yet it can land itself in the dire situation of neglecting the concrete mechanisms that channel the potential for change and on which to build progressive alternatives. With a focus on standards as non-conventional forms of power in contemporary international relations and global political economy, this book has provided ample and widely diverse evidence of the prominence of such mechanisms between macro structural forces and micro individual practices. As Guzzini points out, such focus on modes and mechanisms of global governance reflects a diffusion of power whose pioneer studies go back to the strand of critical realism inaugurated by Strange as well as to Foucault's poststructuralist analysis of 'productive' power and informal rule (Guzzini, 2012). Indeed, Strange, in her call for an 'extensive survey of the extent and limits of non-state authority' considered that the latter 'can only be determined on the basis of outcomes' that she viewed in the shared responsibility with states in the 'complex web of overlapping, symbiotic or conflicting authority in any sector or any who gets what issue'(Strange, 1996: 91, 95, 99). For his part, Foucault was interested in the many ways in which liberalism turns power upside down by taking its distance from the state's coercion and repression capacities, nurturing the suspicion of always governing too much (Foucault, 1994: 820ff). A distinct technology of government is thus needed to induce individuals to create their own disciplinary behaviour. This is basically what prompted Foucault to analyse in great detail the instruments used to translate such disciplinary power in productive forms, that is to say likely to be in conformity with a genuine participation to economic, political, and social orders.

Transposed to present-day concerns regarding the contemporary regime of accumulation, the prominence of such a meso-level of regulation supposes paying attention to what Amin (2004: 226) describes as 'these "small" things ... central in influencing who and what gets ruled in and ruled out or rewarded', and yet rarely featured in analyses of globalisation, global governance, and inequalities. In contrast to one-way analyses focused on the structural and disciplinary power of such 'small things', this book tries to keep a dialectical posture without losing track of the contradictory forces and opposing political economy objectives in situating the power of standards. While large manufacturing and service firms uphold considerable leverage in setting standards in their own interest, it is also at this meso-level that the ambiguous juxtaposition of power instances that play out in transnational hybrid authority can be understood in its ability to provide opportunities for those struggling for progressive change. Spanning all three dimensions of what I referred to as institutional, material, and spatial continuums, such a new topology of hybrid power helps make inroads in the long overdue methodological turn in international relations (Knafo, 2017; Montgomerie, 2017). Here, I deliberately borrow the concept of topology from geometry to convey the idea of plasticity of forms and deformation of structures; the existence of deformable frameworks rules out the possibility of rigid analytical frameworks with the purpose of identifying generalisable propositions. Just as Knafo calls for a renewed perspectivism to cast out dualisms such as 'the opposition between theory (general) and history (specificity), between structures and agency, the global and the local, the national and the international, the state and the market, or more recently the human and nonhumans', understanding in more detail the power of ambiguity across the subjects, objects, and spaces of international relations can be viewed as an alternative strategy 'for gaining perspective on world politics' (Knafo, 2017: 249, 250). To this end, I fully agree with Montgomerie when she claims that sufficient consideration should be given to pluralism as a 'defining feature of the critical school', as long as it describes 'a methodology for investigating capitalism that builds a comprehensive, although no necessarily coherent, understanding from a diversity of corroborating sources' (Montgomerie, 2017: 5, 6).

From this standpoint, a hybrid power based on ambiguity is likely to overcome critiques made on the utopian understanding of post-colonial scholarship focused on the protean and agonistic dimensions of hybridity at the risk of flattening out hierarchical and antagonistic forces in the same political space (Acheraïou, 2011: 153). It also lends support to bringing hierarchy back into Latour's early reading of the social embeddedness of science and technology as a symmetrical anthropology in which 'all the collectives similarly constitute nature' (Latour, 1993: 105, fig. 4.4). Such hybrid power might rather bear some resemblance to the 'politics of Earth' that Latour recently identified in the contradictory movement of both attachment to a soil and detachment made possible by the multiplication of alternatives provided by globalisation (Latour, 2016, 2017). Such emerging yet conflicting rationalities also contribute to what Kessler (2012) identifies as the functional differentiation between distinct temporalities of world society – a temporal conceptualisation largely missing within the confines of this book. Empirically, such a meso-level analysis would look at ambiguity as a strategic resource in the sense of the 'practice turn' in international relations given by everyday and cultural IPE; it would contribute 'to our understanding of the core questions of political economy: the nature of production, trade and finance, the global patterns of distribution and inequality, and the power relations that sustain and constrain them all' (Best and Paterson, 2010: 22). As Best points out, this supposes attentiveness to 'how actors and practices become connected around concrete problems and strategies rather than through predefined fields [as understood by scholars applying Bourdieu's concepts]' (Best, 2014: 24). For instance, any bureaucratic organisation uses ambiguity to devise practices built on lessons drawn from previous failures. Yet it can also become a lever of contestation: 'the very mutability of ambiguity also means that such strategic deployments are always provisional and liable to failure, as a policy can be reinterpreted in unexpected ways' (Best, 2008, 2012: 87). From a broader and more structural perspective, viewing hybrids as a strategic resource based on ambiguity provides insights into the construction of consensus as understood in the Western tradition of Marxism prompted by the writings of Gramsci. Scholarship in international relations has made extensive use of the Gramscian concept of hegemony to emphasise the importance of shared ideas, cultural artefacts, educational programmes, and a flurry of institutional bodies to exert power via a consensus only armoured by coercion. Ambiguity as such looks like a distinctly helpful resource if we go back to Gramsci's own writings. We saw that from a pragmatic posture it provides space for manoeuvre in building compromises; this also applies to concealing the sacrifices made to this end and which Gramsci sees as the required counterpart of hegemony: 'Without doubt, hegemony presumes that we take into account interests and sympathies of groups which hegemony will bear on, that we reach some balanced trade-off, in other words that the leading group makes sacrifices … [which however] cannot concern the most basic points' (Gramsci, 1978: 388, notebook 13, §18 – my translation into English). Overall, this explains why the transnational hybrid authority of instruments such as standards remains, like all hybrids, Janus-faced if it wants to build the slightest hegemonic momentum.

The second area to draw out implications of the argument made in the book is the ambiguity of standards. If ambiguity is a prevailing feature, let alone an ontological attribute, in the ability of standards to create, organise, distribute, and regulate markets and society alike, conventional studies are not enough. Indeed, you cannot just study the economic benefits of standards (DIN, 2000; International Organization for Standardization, 2014), the policy processes in which the institutional supply of standards meets (or not) societal demands for common interest regulation (Spruyt, 2001; Mattli and Woods, 2009), the complementarities between international and domestic institutions (Mattli and Büthe, 2011), or the unrestricted disciplinary power of standards in constituting governmental objects, subjects, and practices (Higgins and Larner, 2010). Laying emphasis on the ambiguous properties of the transnational hybrid authority of standards provides support to underline their intrinsic social contestability, and in turn they may perform opposing political economy objectives and types of relationships between standards and society at large. International standards are unquestionably used as driving forces for broadening the domain and discipline of market self-regulation. As a matter of principle, however, there is no reason to think that they cannot be used as alternative instruments for embedding markets within society. The direction in which the balance will tilt depends, for the most part, on the degree to which society is fairly, substantially, and thoroughly included in standardisation processes; it is also subject to the differentiation of issues likely to be appropriate for such alternative tools of market organisation. This is all the more the case in that standards are not limited to the wide array of topics dealt with by official standardisation bodies such as the ISO but also cover broader sustainability schemes addressing labour and environmental concerns in corporate codes of conduct and multi-stakeholder initiatives. Personal observations drawn from a 'research-action' project devised to support the direct participation of civil society organisations in ISO technical committees have provided evidence that, even with limited resources, it is possible to pool the lay and expert knowledge required for such enhanced participation, to devise strategies likely to support the mobilisation of civil society actors, and eventually exert some (limited) influence over decision making (Hauert et al., 2016; Graz and Hauert, 2019). However, recent scholarship on organic standards shows that many of them keep depoliticising such social contestability and lack international recognition (Fouilleux and Loconto, 2016). A growing literature highlights the complex reality that such supposedly global standards face on the local script of their actual devising procedures and compliance mechanisms (Distelhorst et al., 2015; Bartley and Egels-Zandén, 2016; Mayer et al., 2017; Bartley, 2018).

This argument leads me to distinguish two patterns in the comprehensive topology that spans the institutional, material, and spatial continuums of the transnational hybrid authority of standards. Such patterns signal opposing trends, even if this book provides evidence of some convergence between them. Future developments are likely to face trade-offs between the following opposing forces: the promoters of further socialisation of international standards applied to distinct and explicitly defined topics on the one hand, and, on the other, the advocates of a commodification of technical specification likely to tacitly cut across domains in the same way as management tools impose their discipline to a wide range of working practices. The former will be content with a hefty transfer of the universal scope of law into a catalogue of ad hoc and sectorial standards developed by recognised standard-setting bodies (and some provisions supporting the participation from civil society organisations), backed by intergovernmental rules such as those laid down in the WTO, the EU, and the new generation of mega-trade deals. In contrast, the latter will struggle for worldwide recognition of minimal generic market-based standards, such as quality management and security requirements provided by consultancy firms and consortia competing on the lucrative market of management methodologies and certification as exemplified by the outsourcing and offshoring of business services studied in Chapter 7.

Those two poles oppose trends towards a socialisation of international standards and commodification of technical specifications. Between them, all sorts of variations are likely to span the segments of the institutional, material, and spatial continuums of standardisation. The response made by the American National Standards Institute (ANSI) to the request for comments on the renegotiation of the North American Free Trade Agreement launched by US President Trump in 2017 is a good case in point. ANSI restated its adherence to the principles of the WTO Technical Barriers to Trade (TBT) Agreement guiding the development and implementation of international standards, yet in a way that is likely to reinforce a commodification of technical specifications. According to ANSI, 'Ultimately, the U.S. standardization community supports the fact that there are multiple paths to global relevance – as articulated by the World Trade Organization (WTO) Technical Barriers to Trade (TBT) Agreement – and that it is the marketplace that decides the utility or applicability of any given standard' (American National Standards Institute, 2017b: 1–2). Incidentally, this posture is also consistent with the American tendency to overlook the specificity of services and the distinct types of standards that they may require.

A question remains. As standards are ambiguous but at the same time exert their power by their ability to establish differentiation with opposing political economy objectives, does that power result from their ability to draw boundaries by such differentiation or from being boundary objects blurring varied social orders? My analysis has situated standards along segments of three continuums of a three-dimensional framework in order to appraise the institutional, material, and spatial dimensions which span their authority. As they mix the private and public sphere, physical attributes, and societal values, as well as compliance mechanisms that are both endogenous and exogenous to the principle of territorial sovereignty, standards reflect the idea of 'boundary objects' that Jasanoff uses to describe products of science and technology occupying a 'valued social or moral position precisely because they resist being disambiguated' (Jasanoff, 2005: 27). They create what Winickoff and Mondou (2016: 9) call 'epistemic jurisdictions', whose power stems from their ability to 'produce or warrant technical knowledge for a given political community, topical arena, or geographical territory'. At the same time, as we saw for instance when discussing the Indian office of the world in Chapter 7, standards belong to the 'differentiation strategy' undertaken by firms when competing for markets (Banerjee and Duflo, 2000; Athreye, 2005: 408). As emphasised by Parthasarathy and Srinivasan (2008: 280), standards thus only deliver economic benefits 'by drawing boundaries between those who conform and those who do not'. A third view might be more helpful here: standards *are* and at the same time *draw* boundaries. From this view, the ambiguity of standards is a resource to differentiate markets. The argument made here is akin to the one advanced by Busch. In describing standards as 'recipes for reality', he makes a case for taking full account of the extent to which they 'span the material and the ideal, the positive and the normative, the factual and the ethical, the sacred and the profane', and at the same time are used to differentiate adequately to create new demand for a vast variety of goods and services – what he refers to as 'standardized differentiation' (Busch, 2011: 3, 189). In other words, if standards simultaneously are and draw boundaries, their power rests on their ability to shape thick boundary lines rather than distinct and well-defined epistemic jurisdiction. Just as jurisdictions beset social and political practices that cannot be reduced to the principle of exclusive territoriality of the modern state system, such boundary lines combine enough elements of the opposite poles of their institutional, material, and spatial dimensions to balance the conflicting imperatives of the power of standards.

This brings us to my third point: what are the implications of the power of standards for the rise of services? My syncretic approach has substantially drawn on economic sociology and scholarship inspired by the French *régulation* theory to emphasise the socio-political foundations of standards likely to support the internationalisation of services. According to du Tertre (2008, 2013), two opposite types of institutional outcomes result from the distinct labour relations of services and the potential for standards only lies within one type: the so-called neo-Taylorist option designed to reduce the time needed for establishing the relationship between provider and beneficiary. Such industrial logic involves greater use of machines, information, and communication technology, and stereotypical behaviour. In contrast, standards are ruled out of the opposite strategy of so-called professionalisation, which establishes deontological principles committing providers and beneficiaries alike. My extensive hypothesis emphasising the ability of service standards to meet opposing political economy objectives has led me to explore how standards can apply to both types of institutional outcomes. For instance, management standards, the instruments used in the life insurance market, or the capability business models used in service offshoring in India and elsewhere – all of them clearly rest on an industrial logic. Yet, in other cases, they do not neglect deontological principles that du Tertre sees as limited to the national framework of regulated professions. A case in point is the intense power play in which the European Insurance and Occupational Pensions Authority (EIOPA), the body in charge of the new regulatory power sanctioned by Solvency II, was dragged in to defend the general and consumers' interests against companies' repeated attempts to avoid the implementation of new technical standards resulting in undue increases of their capital buffers. Arguably, another case is the ability of Nasscom, the voice of the IT service industry in India, to turn the Indian service offshoring industry from a standard taker to a standard maker, with the development beginning in 2008 of a new ISO/IEC standard specifically dedicated to India's competitive edge in IT-enabled services and business process outsourcing which was eventually published in 2016 as ISO/IEC 30105.

This book has only made a first foothold on further possible developments with regard to the tensions between industrial and deontological principles underpinning a global service economy. The role of standards in supporting either industrial or deontological principles likely to reinforce worldwide integration of services in domains such as automation, big data, and artificial intelligence will deserve further attention besides concerns as legitimate as those regarding civil liberties, ownership, privacy, or access. In Chapter 7, I showed that even if fully integrated with artificial intelligence, it is unlikely that the future outsourcing of business services could be automated without some form of intermediation that would require a standard to assess quality and security expectations. Similarly, the rise of big data does not rest only on the industrial bedrock to design ever more intricate interoperability standards likely to simultaneously connect billions of devices over the Internet and provide vast power to those in a position to control and exploit such information (Tilley, 2016). Standards can have immense implications beyond technological disputes and market power – and thus pass on industrial and deontological principles alike. They fuel big power policy and ideological rivalry as shown by Chinese early moves to build a rival Internet network or Russian cyberpolitics meddling in electoral processes. As the production and exchange of digital data affects ever more social, cultural, economic, and political spheres, they increasingly belong to what Howard sees as the civilisational choice of Pax Technica – the new era in which we face the challenge of designing the Internet of things in such a way that it may 'either lock us up or set us free'. According to Howard's somehow messianic argument, 'the major battles may no longer be fought by militaries but by corporations with competing technical standards and a vested interest in making systems interoperable or closed' (Howard, 2015: 228, 231). Making big data work for the common good and 'become a robust civic infrastructure' (Howard, 2015: 254) obviously implies a number of preconditions supporting the ability to set standards rather than follow them; but this would lead us too far. Suffice it here to say that while Howard sets progressive potential within the general framework of liberal internationalism and global governance, others situate such potential more explicitly in a post-human ontology viewing a material continuum between the physical and societal materiality of the world. A case in point is Bratton's far-reaching account of what he calls The Stack, a new kind of 'platform sovereignty', in which overlapping layers of a 'standards-based technical-economic system' shape a 'thickened vertical jurisdictional complexity'; against such background of planetary-scale computation, prospects of progressive agency should 'forget human-centered design' and instead carve 'defensible space around the nonhuman *User* [sic] in order to explore the literatures by which human beings can become part of their set' (Bratton, 2015: 7, 42, 4, 288). According to Chandler, big data entails such big promises as 'post-human forms of governance', whose 'empowering and capacity-building relies upon the reconstruction of societies as self-governing, as self-reproducing' (Chandler, 2015: 851, 844).

In a similar vein but more specifically focused on how standards have become ubiquitous in all areas of governance, Busch reckons we have 'some serious (re)thinking about standards to do'. This includes inventing a new form of democracy with an emphasis on deliberative processes and applying a division of powers in the wake of Montesquieu, with its strict separation between bodies setting standards and those enforcing them (certifiers) and still others adjudicating them (accreditors). This would be a precondition to ending the conflicts of interests that plague the current system. It would also require the recognition of standards as 'ontological tools' that bring worlds into being as mobile phones or airports have done, 'so that they can receive the attention they need from various concerned publics *before* they are enacted' (Busch, 2017). With such high stakes, it is surprising that standards likely to shape core socio-technical choices of the future have not aroused more substantial political mobilisation and civic activism. It is less so when we see that standards are still all too commonly regarded just as private voluntary technical specifications geared towards the organisation of markets. I can only hope that this book has made the case for rethinking the power of standards very seriously indeed.

References

Abbott, Kenneth W. and Snidal, Duncan (2001) 'International "standards" and international governance', *Journal of European Public Policy* 8(3): 345–370.

Abbott, Kenneth W. et al. (eds.) (2015) *International Organizations as Orchestrators*, Cambridge: Cambridge University Press.

Acheraïou, Amar (2011) *Questioning Hybridity, Postcolonalism and Globalization*, Houndmills: Palgrave Macmillan.

Acquier, Aurélien and Aggeri, Franck (2008) 'Entrepreneuriat institutionnel et apprentissages collectifs: Le cas de la Global Reporting Initiative (GRI)', *Management International* 12(2): 49–65.

Afnor (2018) *Baromètre international de la normalisation. Position française dans la normalisation internationale*: Association française de normalisation, June.

Aggeri, Franck and Acquier, Aurélien (2008) 'Une généalogie de la pensée managériale sur la RSE', *Revue Française de Gestion* 180: 131–159.

Agnew, John (1994) 'The territorial trap: The geographical assumptions of international relations theory', *Review of International Political Economy* 1 (1): 53–81.

Akerlof, George A. (1970) 'The market for 'lemons': Quality uncertainty and the market mechanism', *The Quarterly Journal of Economics* 84(3): 488–500.

Alemanno, Alberto (2015) 'The Regulatory Cooperation Chapter of the Transatlantic Trade and Investment Partnership: Institutional structures and democratic consequences', *Journal of International Economic Law* 18(3): 625–640.

Allaire, Gilles and Lemeilleur, Sylvaine (2014) 'Normalisation et recherche de garantie : que peut la certification participative?', *The Roles of states and Markets in the Extension of Tripartite Standard Regimes*, Montpelier-CIRAD, 6–7 May.

Allen, John and du Gay, Paul (1994) 'Industry and the rest: The economic identity of services', *Work, Employment & Society* 8(2): 255–271.

Allouche, José, Huault, Isabelle, and Schmidt, Géraldine (2004) 'Responsabilité sociale des entreprises : la mesure détournée?', Actes du 15ème Congrès de l'Association Francophone de Gestion des Ressources Humaines AGRH, Montreal.

Amable, Bruno (2000) 'Institutional complementarity and diversity of social systems of innovation and production', *Review of International Political Economy* 7: 645–687.

American National Standards Institute (2016) *United States Standards Strategy*, New York.

(2017a) *G20 Leaders Statement Includes Reference to Industry-Led International Standards*, 10 July. Online.

(2017b) *ANSI Response to Request for Comments: Modernization of the North American Free Trade Agreement (NAFTA)*, Washington DC: Office of the United States Trade Representative (USTR), 28 June, FR Doc. 2017–10603; USTR-2017–0006.

Amin, Ash (2004) 'Regulating economic globalization', *Transactions of the Institute of British Geographers* 29(2): 217–233.

Aoki, Masahiko (1994) 'The contingent governance of teams: Analysis of institutional complementarity', *International Economic Review* 35(3): 657–676.

Aradau, Claudia, Lobo-Guerrero, Luis, and van Munster, Rens (2008) 'Security, Technologies of Risk, and the Political: Guest Editors' Introduction', *Security Dialogue* 39(2/3): 147–154.

ASTM International (n.d.) *White Paper: Strengthening Transatlantic Trade through a Common Understanding on Standards*, available online at: www.astm.org/MEETINGS/images/SES_ASTMWhitePaper.pdf.

Athreye, Suma S. (2005) 'The Indian software industry and its evolving service capability', *Industrial and Corporate Change* 14(3): 393–418.

Avant, Deborah D., Finnemore, Martha, and Sell, Susan K. (eds.) (2010) *Who Governs the Globe?* Cambridge: Cambridge University Press.

Bair, Jennifer (2017) 'Contextualising compliance: Hybrid governance in global value chains', *New Political Economy* 22(2): 169–185.

Baker, Tom and Simon, Jonathan (eds.) (2002) *Embracing Risk: The Changing Culture of Insurance and Responsibility*, Chicago: The University of Chicago Press.

Bakhtin, Mikhail Mikhailovich (1981) 'Discourse in the Novel', in Holquist, Michael (ed.) *The Dialogic Imagination: Four Essays*, Austin: University of Texas Press, 259–422.

Banerjee, Abhijit Vinayak and Duflo, Esther (2000) 'Reputation effects and the limits of contracting: A study of the Indian software industry', *Quarterly Journal of Economics* 115(3): 989–1017.

Barnett, Michael and Duvall, Raymond (eds.) (2005) *Power in Global Governance*, Cambridge: Cambridge University Press.

Barras, Richard (1986) 'Towards a theory of innovation in services', *Research Policy* 15: 161–173.

Barraud de Lagerie, Pauline (2011) 'Quand l'efficacité anticipée d'une institution entrave sa pleine genèse: La construction de la norme ISO 26000 sur la responsabilité sociale', in François, Pierre (ed.) *Vie et mort des institutions marchandes*, Paris: Presses de Sciences Po, 169–194.

Barry, Andrew (2013) 'The translation zone: Between actor-network theory and international relations', *Millennium – Journal of International Studies* 41(3): 413–429.

Bartley, Tim (2011) 'Transnational governance as the layering of rules: Intersections of public and private standards', *Theoretical Enquiries in Law* 12: 517–542.

(2018) *Rules without Rights: Land, Labor, and Private Authority in the Global Economy*, Oxford: Oxford University Press.

Bartley, Tim and Egels-Zandén, Niklas (2016) 'Beyond decoupling: Unions and the leveraging of corporate social responsibility in Indonesia', *Socio-Economic Review* 14(2): 231–255.

Baumol, William J. (1967) 'Macroeconomics of unbalanced growth: The anatomy of the urban crisis', *American Economic Review* 57: 415–426.

Beck, Ulrich (1992) *Risk Society: Towards a New Modernity*, London: SAGE.

(2000) 'Risk Society Revisited: Theory, Politics, and Research Programmes', in Adam, Barbara, Beck, Ulrich and van Loon, Joost (eds.) *The Risk Society And Beyond: Critical Issues for Social Theory*, London: Sage, 211–229.

Bell, Daniel (1973) *The Coming of Post-Industrial Society; A Venture in Social Forecasting*, New York: Basic Books.

Belloni, Roberto (2012) 'Hybrid peace governance', *Global Governance* 18(1): 21–38.

Berens, Jasper (2015) 'Why the UK annuity market won't go the way of the dinosaurs': J.P. Morgan Asset Management, 24 February.

Bergström, Sven (2004) 'Dix années d'activités des syndicats suédois dans le processus de normalisation national et européen', *Newsletter du BTS* 24–25: 7–12.

Berle, Adolf and Means, Gardiner (1932) *Modern Corporation and Private Property*, New York: Commerce Clearing House.

Best, Jacqueline (2005) *The Limits of Transparency: Ambiguity in the History of International Finance*, Ithaca, NY: Cornell University Press.

(2008) 'Ambiguity, uncertainty and risk : Rethinking indeterminacy', *International Political Sociology* 2(4): 355–374.

(2012) 'Bureaucratic ambiguity', *Economy and Society* 41(1): 84–106.

(2013) *Governing Failure: Provisional Expertise and the Transformation of Global Development Finance*, Cambridge: Cambridge University Press.

Best, Jacqueline and Paterson, Matthew (eds.) (2010) *Cultural Political Economy*, London: Routledge.

Best, Jacqueline and Walters, William (2013) '"Actor-network theory" and international relationality: Lost (and found) in translation', *International Political Sociology* 7(3): 332–334.

Bhabha, Homi K. (1994) *The Location of Culture*, London: Routledge.

Blake, David et al. (2013) 'The new life market', *The Journal of Risk and Insurance* 80(3): 501–558.

Blake, David, Boardman, Tom, and Cairns, Andrew (2014) 'Sharing longevity risk: Why governments should issue longevity bonds', *North American Actuarial Journal* 18(1): 258–277.

Blind, Knut (2003) *Standards in the Service Sector: An Explorative Study*: European Commission, Enterprise Directorate-General, April.

(2004) *The Economics of Standards: Theory, Evidence, Policy*, Cheltenham: Edward Elgar.

Blind, Knut, Petersen, Sören, and Riillo, Cesare (2017) 'The Impact of Standards and Regulation on Innovation in Uncertain Markets', *Research Policy* 46(1): 249–264.

Blinder, Alan S. (2006) 'Offshoring: The Next Industrial Revolution?', *Foreign Affairs* 85(2): 113–128.

Boden, Mark and Miles, Ian (2000a) 'Conclusions: Beyond the Services Economy', in Boden, Mark and Miles, Ian (eds.) *Services and the Knowledge-Based Economy*, London: Continuum, 247–264.

(eds.) (2000b) *Services and the Knowledge-Based Economy*, London: Continuum.

Borraz, Olivier (2007) 'Governing standards: The rise of standardization processes in France and in the EU', *Governance* 20(1): 57–84.

Brassett, James and Tsingou, Eleni (2011) 'The politics of legitimate global governance', *Review of International Political Economy* 18(1): 1–16.

Bratton, Benjamin H. (2015) *The Stack: On Software and Sovereignty*, Cambridge, MA: MIT Press.

Brémond, Mireille (1996) '*La Machine infernale* de J. Cocteau: Réflexion sur l'hybridité', *Uranie* 6: 144–156.

Brette, Olivier and Moriset, Bruno (2009) 'Bringing down territorial inequalities in the digital economy: An evolutionary institutional approach', *Journal of Economic Issues* 43(2): 495–502.

Brown, Jeffrey R. et al. (eds.) (2001) *The Role of Annuity Markets in Financing Retirement*, Cambridge, MA: MIT Press.

Brunsson, Nils, Jacobsson, Bengt et al. (2000) *A World of Standards*, Oxford: Oxford University Press.

Brunsson, Nils, Rasche, Andreas, and Seidl, David (2012) 'The dynamics of standardization: Three perspectives on standards in organization studies', *Organization Studies* 33(5–6): 613–632.

Bryson, John R. and Daniels, Peter W. (2007a) 'World of Services: From Local Service Economies to Offshoring or Global Sourcing', in Bryson, John R. and Daniel, Peter W. (eds.) *The Handbook of Service Industries*, Cheltenham: Edward Elgar, 1–16.

(eds.) (2007b) *The Handbook of Service Industries*, Cheltenham: Edward Elgar.

Busch, Lawrence (2011) *Standards. Recipes for Reality*, Cambridge, MA: MIT Press.

(2017) 'Standards and their problems', *Research in Rural Sociology and Development* 24: 97–114.

Büthe, Tim (2010) 'Private regulation in the global economy: A (p)review', *Business and Politics* 12(3): 1–38.

Cai, Phoenix X. F. (2016) 'Regulatory coherence and standardization mechanisms in the Trans-Pacific Partnership', *British Journal of American Legal Studies* 5(2): 505–538.

Callon, Michel, Lascoumes, Pierre, and Barthe, Yannick (2001) *Agir dans un monde incertain: Essai sur la démocratie technique*, Paris: Seuil.

Callon, Michel, Méadel, Cécile, and Rabeharisoa, Vololona (2002) 'The economy of qualities', *Economy and Society* 31(2): 194–217.

Cameron, Angus and Palan, Ronen (2003) *The Imagined Economies of Globalisation*, London: Sage.

Capron, Michel and Quairel-Lanoizelée, Françoise (2007) *La responsabilité sociale d'entreprise*, Paris: La Découverte.

Capron, Michel, Quairel-Lanoizelée, Françoise, and Turcotte, Marie-France (eds.) (2010) *Iso 26000 : une Norme "hors norme" ?: Vers une conception mondiale de la responsabilité sociétale*, Paris: L'Harmattan.

Cass Business School and Hymans Robertson LLP (2014) *Longevity Basis Risk. A Methodology for Assessing Basis Risk*, London: The Institute and Faculty of Actuaries and the Life and Longevity Markets Association, December.

Cassese, Sabino (2005) 'Administrative law without the state? The challenge of global regulation', *New York University Journal of International Law and Politics* 37(4): 663–694.

Castells, Manuel (2001) *La société en réseaux: L'ére de l'information*. 2nd edn, Paris: Fayard.

CEN (2017) *Strategic Plan on Services Standardization to Implement the Ambition 2020*, Brussels: European Committee for Standardization.

Cernat, Lucian and Kutlina-Dimitrova, Zornitsa (2014) *Thinking in a box : a 'mode 5' approach to service trade*, European Commission – Chief Economist Note, Working Paper No 1, Brussels.

Chamberlin, Edward Hastings (1925) *The Theory of Monopolistic Competition*, Cambridge, MA: Harvard University Press.

Chandler, David (2015) 'A World without causation: Big data and the coming of age of posthumanism', *Millennium – Journal of International Studies* 43(3): 833–851.

Chanteau, Jean-Pierre (2011) 'L'économie de la responsabilité sociétale d'entreprise (RSE) : éléments de méthode institutionnaliste', *Revue de la régulation* 9(1).

CHESSS Consortium (2009) *Consolidated Report on the Feasibility Study in Response to EU Mandate M/ 371*: CEN's Horizontal European Services Standardization Strategy.

Churchill, Craig and McCord, Michael J. (2012) 'Current Trends in Microinsurance', in Churchill, Craig and Matul, Michal (eds.) *Protecting the Poor. A Microinsurance Compendium*, Volume II, Geneva: International Labour Organisation, 8–39.

Clark, Geoffrey and Anderson, Gregory (2010) 'Introduction', in Clark, Geoffrey et al. (eds.) *The Appeal of Insurance*, Toronto: University of Toronto Press, 3–15.

Clark, Geoffrey et al. (2010) *The Appeal of Insurance*, Toronto: University of Toronto Press.

Colàs, Alejandro (2002) *International Civil Society: Social Movements in World Politics*, Cambridge: Polity.

Collier, Stephen J. (2008) 'Enacting catastrophe: Preparedness, insurance, budgetary rationalization', *Economy and Society* 37(2): 224–250.

Commission on Global Governance (1995) *Our Global Neighbourhood: The Report of the Commission on Global Governance*, Oxford: Oxford University Press.

Commons, John R. (1934) *Institutional Economics: Its Place in Political Economy*, New York: Macmillan.

Copeland, B. R. and Mattoo, A. (2008) 'The Basic Economics of Services Trade', in Mattoo, Aaditya, Stern, Robert Mitchell, and Zanini, Gianni (eds.) *Handbook of International Trade in Services*, Oxford: Oxford University Press, 84–129.

Coughlan, Guy et al. (2007) *q-Forwards: Derivatives for Transferring Longevity and Mortality Risks*, London: J.P. Morgan Pension Advisory Group, July, www.jpmorgan.com/cm/cs?pagename=JPM/DirectDoc&urlname=LM_Q_forwards.pdf&track=no, accessed 12 April 2015.

Council of the European Union (2016) *Joint Interpretative Instrument on the Comprehensive Economic and Trade Agreement (CETA) between Canada and the European Union and its Member States.* 27 October 2016, 13541/16, Brussels.

Cox, Robert W. (1981) 'Social forces, states and world orders: Beyond international relations theory', *Millennium – Journal of International Studies* 10(2): 126–155.

(1987) *Production, Power and World Order: Social Forces in the Making of History*, New York: Columbia University Press.

(2002) *The Political Economy of a Plural World: Critical Reflections on Power, Morals and Civilization*, with Schechter, Michael G., London: Routledge.

Cutler, A. Claire (1999) 'Private Authority in International Trade Relations: The Case of Maritime Transport', in Cutler, A. Claire, Haufler, Virginia, and Porter, Tony (eds.) *Private Authority and International Relations*, Albany, NY: SUNY Press, 283–329.

(2003) *Private Power and Global Authority: Transnational Merchant Law in the Global Political Economy*, Cambridge: Cambridge University Press.

(2010) 'The legitimacy of private transnational governance: experts and the transnational market for force', *Socio-Economic Review* 8(1): 157–185.

Cutler, A. Claire, Haufler, Virginia, and Porter, Tony (eds.) (1999) *Private Authority and International Affairs*, Albany, NY: SUNY Press.

Czaya, Axel and Hesser, Wilfried (2001) 'Standardization systems as indicators of mental, cultural and socio-economic states', *Knowledge, Technology, & Policy* 14(3): 24–40.

Daniels, Peter W. and Bryson, John R. (2002) 'Manufacturing services and servicing manufacturing: Changing forms of production in advanced capitalist economies', *Urban studies* 39(5–6): 977–991.

Danielsson, Jon et al. (2012) 'Countercyclical regulation in Solvency II: Merits and flaws', *VOX CEPR's Policy Portal*, 23 June, accessed 14 April 2015.

Das, Diya and Dharwadkar, Ravi (2009) 'Cultural Mimicry and Hybridity: On the Work of Identity in International Call Centers in India', in Banerjee, Subhabrata Bobby, Chio, Vanessa C. M., and Mir, Raza (eds.) *Organizations, Markets and Imperial Formations. Towards an Anthropology of Globalisation*, Cheltenham: Edward Elgar.

David, Paul A. (1985) 'Clio and the economics of QWERTY', *The American Economic Review* 75(2): 332–337.

Davis, George and Garda, Giovanni (2009) 'ACORD: a way forward in seamless exposure data exchange', *AIR Currents*, 11 August, www.air-worldwide.com/Publications/AIR-Currents/ACORD–A-Way-Forward-in-Seamless-Exposure-Data-Exchange/, accessed 30 July 2015.

De Ville, Ferdi and Siles-Brugge, Gabriel (2016) *TTIP: The Truth about the Transatlantic Trade and Investment Partnership*, London: Polity Press.

Deming, William E. (1981) *Management of Statistical Techniques for Quality and Productivity*, New York: New York University Press.

Dezalay, Yves and Garth, Bryant G. (2011) 'Hegemonic battles, professional rivalries, and the international division of labor in the market for the import and export of state-governing expertise', *International Political Sociology* 5(3): 276–293.

Dicken, Peter (2015) *Global Shift: Mapping the Changing Contours of the World Economy*, 7th edn., London: SAGE.

DIN (2000) *The Economic Benefits of Standardization*, Berlin: Beuth Verlag.

Dingwerth, Klaus and Eichinger, Margot (2010) 'Tamed transparency: How information disclosure under the Global Reporting Initiative fails to empower', *Global Environmental Politics* 10(3): 74–96.

Distelhorst, Greg et al. (2015) 'Production goes global, compliance stays local: Private regulation in the global electronics industry', *Regulation & Governance* 9(3): 224–242.

Djelic, Marie-Laure and den Hond, Frank (2014) 'Symposium on multiplicity and plurality in the world of standards', *Business and Politics* 16(1): 67–77.

Djelic, Marie-Laure and Quack, Sigrid (eds.) (2010) *Transnational Communities: Shaping Global Economic Governance*, Cambridge: Cambridge University Press.

Djelic, Marie-Laure and Sahlin-Andersson, Kerstin (eds.) (2006) *Transnational Governance: Institutional Dynamics of Regulation*, Cambridge: Cambridge University Press.

Djellal, Farida and Gallouj, Faïz (eds.) (2002) *Nouvelle économie des services et innovation*, Paris: L'Harmattan.

(2010) 'Services, innovation and performance: General presentation', *Journal of Innovation Economics & Management* 5(1): 5–15.

Dopfer, Kurt (ed.) (2005) *The Evolutionary Foundation of Economics*, Cambridge: Cambridge University Press.

Dossani, Rafiq and Kenney, Martin (2007) 'The next wave of globalization: Relocating service provision to India', *World Development* 35(5): 772–791.

Doyle, Aaron (2011) 'Introduction: Insurance and business ethics', *Journal of Business Ethics* 103(1): 1–5.

Doyle, Aaron and Ericson, Richard (2010) 'Five Ironies of Insurance', in Clark, Geoffrey et al. (eds.) *The Appeal of Insurance*, Toronto: University of Toronto Press, 226–247.

Du Tertre, Christian (2002) 'Activités immatérielles et relationnelles: quels nouveaux enjeux de régulaton pour les secteurs et les territoires?', *Géographie, Economie, Société* 4(2): 181–204.

(2008) 'Globalisation et services: pour un changement de perspectives', in Lyon-Caen, A. and Urban, Q. (eds.) *Le droit du travail à l'épreuve de la globalisation*, Paris: Dalloz, 61–81.

(2013) 'Configurations productives de services et internationalisation: une approche régulationniste', in Graz, Jean-Christophe and Niang, Nafi (eds.) *Services sans frontières. Mondialisation, normalistion, et régulation de l'économie des services*, Paris: Presses de Science Po.

Dudouet, François-Xavier , Mercier, Delphine, and Vion, Antoine (2006) 'Politiques internationales de normalisation: Quelques jalons pour la recherche empirique', *Revue Française de Science Politique* 56(3): 367–392.

Eaton, Monica et al. (2013) *The CoRe NTMs Database Version 2: A Compilation of Reported Non-Tariff Measures*, Office of Economics Working Paper/U.S. International Trade Commission, Working Paper No 2013–01A, Washington DC.

Eckstein, Harry (1975) 'Case Studies and Theory in Political Science', in Greenstein, Fred I. and Polsby, Nelson W. (eds.) *Handbook of Political Science. Vol. 7, Political Science: Scope and Theory*, Reading, MA: Addison-Wesley, 79–138.

Egan, Michelle (2001) *Constructing a European Market: Standards, Regulation, and Governance*, Oxford: Oxford University Press.

Egyedi, Tineke (2005) 'Beyond Consortia, Beyond Standardization? Redefining the Consortium Problem', in Jacobs, Kai (ed.) *Advanced Topics in Information Technology Standards and Standardization Research*, London: IGI publishing, 91–110.

Elam, Mark (1999) 'Living dangerously with Bruno Latour in a hybrid world', *Theory, Culture & Society* 16(4): 1–24.

Enderwick, Peter (2007) 'Multinational Service Firms and Global Strategy', in Bryson, John R. and Daniels, Peter W. (eds.) *The Handbook of Service Industries*, Cheltenham: Edward Elgar, 258–274.

Ericson, Richard, Doyle, Aaron, and Barry, Dean (2003) *Insurance as Governance*, Toronto: University of Toronto Press.

European Commission (2016a) *Annex to the Proposal for a Council Decision on the signing on behalf of the European Union of the Comprehensive Economic and Trade Agreement between Canada of the one part, and the European Union and its Member States, of the other part*, Strasbourg, 5 July, COM(2016) 444 final – Annex 1.

(2016b) *The Transatlantic Trade and Investment Partnership (TTIP). The Top 10 Myths about TTIP. Separating Fact from Fiction*, Luxembourg.

(2016c) *Standardisation Package. Commission Staff Working Document. Tapping the Potential of European Service Standards to Help Europe's Consumers and Businesses*, Brussels, 1 June, SWD(2016) 186 final.

(2016d) *Report from the Commission to the European Parliament and the Council on the Implementation of the Regulation (EU) No 1025/2012 from 2013 to 2015*, Brussels, 1 June, COM(2016) 212 final.

(2016e) *Standardisation package. European Standards for the 21st Century. Communication from the Commission to the European Parliament, the Council, the European Social and Economic Committee and the Committee of the Regions*, 1 June, COM(2016) 358 final.

European Parliament (2014) *Hearing of Cecilia Malmström Commissioner-Designated (Trade), Verbatim Report, Monday, 29 September 2014*, Brussels.

Evans, Peter (1995) *Embedded Autonomy. States and Industrial Transformation*, Princeton, NJ: Princeton University Press.

Evans, Ross, Hegeman, Neal, and Penn, Emily (2013) 'The ramifications of extrapolations', *Solvency II Wire*, London, 24 April, www.solvencyiiwire.com/the-ramifications-of-extrapolations/8372925, accessed June 2015.

EXPRESS (2010) *Standardization for a Competitive and Innovative Europe: A Vision for 2020*: The Expert Panel for the Review of the European Standardisation System, available online at: www.anec.eu/attachments/Definitive EXPRESS report.pdf.

Feenberg, Andrew (1991) *Critical Theory of Technology*, New York: Oxford University Press.

Fernie, Sue and Metcalf, David (1998) *(Not) Hanging on the Telephone: Payment Systems in the New Sweatshops*, Centre for Economic Performance, Working Paper No 390, London.

Foucault, Michel (1994) *Dits et écrits – 1954–1988*. Vol. III - 1976–1979, Paris: Gallimard.

Fouilleux, Eve and Loconto, Allison (2016) 'Voluntary standards, certification, and accreditation in the global organic agriculture field: a tripartite model of techno-politics', *Agriculture and Human Values*: 1–14 (published online: 8 February).

Friedman, Thomas L. (2006) *The World is Flat: The Globalized World in the Twenty-First Century*, London: Penguin.

Frydman, Benoît (2014) 'Prendre les standards et les indicateurs au sérieux', in Frydman, Benoît and Wayenberger, Arnaud Van (eds.) *Gouverner par les standards et les indicateurs*, Brussels: Bruylant.

Gadrey, Jean (2003) *Socio-économie des services, "Repères"*, Paris: La Découverte.

(2010) *Adieu à la croissance: Bien vivre dans un monde solidaire*, Paris: Les petits matin & Alternatives économiques.

Garson, Barbara (1988) *The Electronic Sweatshop: How Computers are Transforming the Office of the Future*, New York: Simon & Shuster.

Gibbon, Peter and Henriksen, Lasse Folke (2011) 'On the Pre-history of ISO 9000: The Making of a Neo-liberal Standard', in Ponte, Stefano, Gibbon, Peter, and Vestergaard, Jakob (eds.) *Governing through Standards*, Houndmills: Palgrave Macmillan, 130–158.

Gill, Stephen and Cutler, A. Claire (2015) *New Constitutionalism and World Order*, Cambridge: Cambridge University Press.

Godin, Christian (1996) 'L'hybride entre la puissance et l'effroi', *Uranie* 6: 37–47.

Gramsci, Antonio (1978) *Cahiers de prison: Cahiers 10, 11, 12 et 13*, Paris: Gallimard.

Grande, Edgar and Pauly, Louis W. (eds.) (2005) *Complex Sovereignty: Reconstituting Political Authority in the Twenty-First Century*, Toronto: University of Toronto Press.

Graz, Jean-Christophe (2008) 'Gare aux hybrides: Mythes et réalités de la gouvernance de la mondialisation', *Etudes Internationales XXXIX* (3): 361–385.

(2018) 'Global Corporations and the Governance of Standards', in Nölke, Andreas and May, Christian (eds.) *Handbook of the International Political Economy of the Corporation*, Aldershot: Edward Elgar, 448–461.

Graz, Jean-Christophe and Hauert, Christophe (2014) 'Beyond the transatlantic divide: The multiple authorities of standards in the global political economy of services', *Business and Politics* 16(1): 113–150.

(2019) 'Translating technical diplomacy: The participation of civil society organisations in international standardisation', *Global Society* 32(2): 163–183.

Graz, Jean-Christophe and Nölke, Andreas (eds.) (2008) *Transnational Private Governance and its Limits*, London: Routledge.

Green, Jessica F. (2014) *Rethinking Private Authority: Agents and Entrepreneurs in Global Environmental Governance*, Princeton, NJ: Princeton University Press.

Grönroos, Christian (1990) *Service Management and Marketing Managing the Moments of Truth in Service Competition*, Lexington, MA: Lexington Books.

Grove, Kevin J. (2010) 'Insuring "our common future?" Dangerous climate change and the biopolitics of environmental security', *Geopolitics* 15(3): 536–563.

Guler, Islin, Guillén, Mauro F., and Macpherson, John Muir (2002) 'Global competition, institutions, and the diffusion of organizational practices: The international spread of ISO 9000 quality certificates', *Administrative Science Quarterly* 47: 207–232.

Gusman, Phil (2010) 'Are Insurers Too Dependent On Catastrophe Modeling? Vendors emphasize that predictive systems are tools, with uncertainties inevitable', *National Underwriter Property & Casualty* 22 February.

Gutner, Tamar (2010) 'When 'Doing Good' Does Not: The IMF and the Millennium Development Goals', in Avant, Deborah D., Finnemore, Martha, and Sell, Susan K. (eds.) *Who Governs the Globe?*, Cambridge: Cambridge University Press, 266–291.

Guzzini, Stefano (2012) 'The Ambivalent "Diffusion of Power" in Global Governance', in Guzzini, Stefano and Neumann, Iver B. (eds.) *The Diffusion of Power: International Political Economy Meets Foucault*, London: Palgrave Macmillan, 1–37.

Guzzini, Stefano and Neumann, Iver B. (eds.) (2012) *The Diffusion of Power: International Political Economy Meets Foucault*, London: Palgrave Macmillan.

Hall, Debra J. and Berset, Jérôme (2010) 'Toward Global Reinsurance Regulation', in Kempler, Cecelia et al. (eds.) *Global Perspectives on Insurance Today*, London: Palgrave Macmillan, 91–108.

Hall, Rodney Bruce and Biersteker, Thomas J. (2002a) 'The Emergence of Private Authority in in the International System', in Hall, Rodney Bruce and Biersteker, Thomas J. (eds.) *The Emergence of Private Authority in Global Governance*, Cambridge: Cambridge University Press, 3–22.

(eds.) (2002b) *The Emergence of Private Authority in Global Governance*, Cambridge: Cambridge University Press.

Hanna, Nagy (1994) *Exploiting Information Technology for Development: A Case Study of India*, Washington DC: The World Bank Group.

Hartmann, Eva (2013) 'Tous pareils, tous différents ? Les chaînes d'équivalence dans l'internationalisation des services professionnels', in Graz, Jean-Christophe and Niang, Nafi (eds.) *Services sans frontières: mondialisation, normalisation et régulation de l'économie des services*, Paris: Presses de Science Po, 187–222.

Hatzopoulos, Vassilis (2012) 'Forms of Mutual Recognition in the Field of Services', in Lianos, Ioannis and Odudu, Okeoghene (eds.) *Regulating Trade in Services in the EU and the WTO*, Cambridge: Cambridge University Press, 59–98.

Hauert, Christophe (2014) *La participation des consommateurs à la construction de l'autorité des normes internationales: vers une économie politique internationale de la traduction*, Thèse de Doctorat, présentée à la Faculté des sciences sociales et politiques, pour l'obtention du grade de Docteur ès science politique, Université de Lausanne.

Hauert, Christophe et al. (2016) 'Les arènes de la normalisation internationale à l'épreuve de la participation : le projet INTERNORM', *Participations* 1(14): 207–235.

Haufler, Virginia (1997) *Dangerous Commerce: Insurance and the Management of International Risk*, Ithaca, NY: Cornell University Press.

(2009) 'Insurance and Reinsurance in a Changing Climate', in Selin, Henrik and VanDeveer, Stacy D. (eds.) *Changing Climates in North American Politics*, Cambridge, MA: MIT Press, 241–262.

Hawkins, Richard, Blind, Knut, and Page, Robert (eds.) (2017) *Handbook of Innovation and Standards*, Aldershot: Edward Elgar.

Heeks, Richard (1996) *India's Software Industry: State Policy, Liberalisation and Industrial Development*, New Delhi: Sage.

Heimer, Carol A. (2002) 'Insuring More, Ensuring Less: The Costs and Benefits of Private Regulation through Insurance', in Baker, Tom and Simon, Jonathan (eds.) *Embracing Risk. The Changing Culture of Insurance and Responsibility*, Chicago: The University of Chicago Press, 116–145.

Held, David et al. (1999) *Global Transformation: Politics, Economics and Culture*, Cambridge: Polity Press.

Helleiner, Gerald K. (2001) 'Markets, politics, and globalization: Can the global economy be civilized?', *Global Governance* 7(3): 243–263.

Hewitt de Alcantara, Cynthia (1998) 'Du bon usage du concept de gouvernance', *Revue Internationale des Sciences Sociales* 50(1): 109–118.

Hewson, Martin and Sinclair, Timothy J. (1999) 'The Emergence of Global Governance Theory', in Hewson, Martin and Sinclair, Timothy J. (eds.) *Approaches to Global Governance Theory*, Albany, NY: SUNY Press, 3–22.

Hibou, Béatrice (2016) 'Neolibral Democracy as an Expression of Hybrid Rule', in Hurt, Shelley and Lipschutz, Ronnie (eds.) *Hybrid Rule and State Formation: Private-Public Power in the 21st Century*, London: Routledge, 59–78.

Higgins, Vaughan and Larner, Wendy (eds.) (2010) *Calculating the Social: Standards and the Reconfiguration of Governing*, London: Palgrave Macmillan.

Higgins, Winton and Tamm Hallström, Kristina (2007) 'Standardization, globalization and the rationalities of government', *Organization* 14(5): 685–704.

Higgott, Richard, Underhill, Geoffrey, and Bieler, Andreas (eds.) (1999) *Non-State Actors and Authority in the Global System*, London and New York: Routledge.

Hill, Peter (1977) 'On goods and services', *Review of Income and Wealth* 23(4): 315–338.

Hirschman, Albert O. (1970) *Exit, Voice, and Loyalty: Responses to Decline in Firms, Organizations, and States*, Cambridge, MA: Harvard University Press.

Hix, Simon (1998) 'The study of the European Union II: The "New Governance" agenda and its rival', *Journal of European Public Policy* 5(1): 38–65.

Hogan, Michael J. (1977) *Informal Entente: The Private Structure of Cooperation in Anglo-American Economic Diplomacy, 1918–1928*, Columbia: University of Missouri Press.

Holman, Otto (2004) 'Asymmetrical regulation and multidimensional governance in the European Union', *Review of International Political Economy* 11(4): 714–735.

Howard, Philip N. (2015) *Pax Technica: How the Internet of Things May Set Us Free or Lock Us Up*, New Haven, CT: Yale University Press.

Humphrey, Christopher and Loft, Anne (2011) 'Moving Beyond Nuts and Bolts: The Complexities of Governing a Global Profession through International Standards', in Ponte, Stefano, Gibbon, Peter, and Vestergaard, Jakob (eds.) *Governing through Standards*, Houndmills: Palgrave Macmillan: 102–129.

Hurt, Shelley and Lipschutz, Ronnie (eds.) (2016) *Hybrid Rule and State Formation: Public-Private Power in the 21st Century*, London: Routledge.

Hymans Robertson (2018) 'Buy outs, buy-ins and longevity hedging – H2 2017', London: Hymans Robertson LPP, 23 March 2018.

Illeris, Sven (2007) 'The Nature of Services', in Bryson, John R. and Daniels, Peter W. (eds.) *The Handbook of Service Industries*, Cheltenham: Edward Elgar, 19–33.

Inkmann, Joachim, Lopes, Paula, and Michaelides, Alexander (2011) 'How Deep is the annuity market participation puzzle?', *The Review of Financial Studies* 24(1): 279–319.

International Association of Insurance Supervisors (2014) *Risk-based Global Insurance Capital Standard: Public Consultation Document Comments due by 16 February 2015*: IAIS, 17 December.

(2015) 'The IAIS Risk-based Global Insurance Capital Standard (ICS): Ultimate and Interim Goals, Principles for Development and Delivery Process, IAIS, updated 25 June 2015', available online at: http://iaisweb.org/index.cfm?event=getPage&nodeId=25229, accessed 20 July 2015.

International Institute for Strategic Studies (2016) 'Chapter ten: Country comparisons – commitments, force levels and economics', *The Military Balance* 116(1): 481–492.

International Monetary Fund (2015) *United States: Financial Sector Assessment Program-Detailed Assessment of Observance on the Insurance Core Principles*, Washington DC: IMF, April 02.

International Organization for Standardization (2014) *The Economic Benefits of Standardization*, Geneva: ISO.

(1995) *ISO Bulletin* 26(1).

(2016a) *Strategy for Service Standardization*, Geneva: ISO.

(2016b) *ISO Strategy for Services – Presentation*, Geneva: ISO.

(2017) *ISO Workshop Report. Growing the Services Industry with ISO Standards. 28 April, Vancouver, Canada*, Geneva: ISO.

(2016) 'The service economy', *ISO Focus* (116).

Jacquet, Pierre, Pisani-Ferry, Jean, and Tubiana, Laurence (2002) *Gouvernance Mondiale: Rapport de Synthèse*, Paris: La Documentation Française/Conseil d'Analyse Economique.

Jasanoff, Sheila (2005) *Design on Nature: Science and Democracy in Europe and the United States*, Princeton, NJ: Princeton University Press.

(ed.) (2004) *States of Knowledge : The Co-Production of Science and Social Order*, New York: Routledge.

Jegen, Maya and Mérand, Frédéric (2014) 'Constructive ambiguity: Comparing the EU's energy and defence policies', *West European Politics* 37(1): 182–203.

Jensen, J. Bradford (2011) *Global Trade in Services – Fear, Facts, and Offshoring*, Washington DC: Peterson Institute for International Economics.

JISC (2013) *Japan's Standardization Policy 2013*, Tokyo: Japanese Industrial Standards Committee, Ministry of Economy Trade and Industry, March.

Johnson, Michael and Nilsson, Lars (2003) 'The importance of reliability and customization from goods to services', *The Quality Management Journal* 10 (1): 8–19.

Jurcys, Paulius, Kjaer, Poul, and Yatsunami, Ren (eds.) (2013) *Regulatory Hybridization in the Transnational Sphere*, Leiden: Martinus Nijhoff.

Justice, Cliff (2012) 'The Death of Outsourcing', *Shared Services and Outsourcing Advisory*, KPMG International, June.

Kapur, D. (2002) 'The causes and consequences of India's IT boom', *India Review* 1(2): 91–110.

Karpik, Lucien (1989) 'L'économie de la qualité', *Revue française de sociologie* 30 (2): 187–210.

Keohane, Robert O. (2011) 'Global governance and legitimacy', *Review of International Political Economy* 18(1): 99–109.

Kessler, Oliver (2012) 'World society, social differentiation and time', *International Political Sociology* 6: 77–94.

Kingsbury, Benedict, Krisch, Nico, and Stewart, Richard B. (2005) 'The emergence of global administrative law', *Law and Contemporary Problems* 68(3/4): 15–61.

Kjaer, Poul (2013) 'Introduction', in Jurčys, Paulius, Kjaer, Poul, and Yatsunami, Ren (eds.) *Regulatory Hybridization in the Transnational Sphere*, Leiden: Martinus Nijhoff, 1–10.

Knafo, Samuel (2017) 'A Methodological Turn Long Overdue: Or, Why It Is Time for Critical Scholars to Cut Their Losses', in Dyvik, Synne L., Selby, Jan, and Wilkinson, Rorden (eds.) *What's the Point of International Relations?*, London: Routledge, 241–252.

Knight, Frank H. (1971) *Risk, Uncertainty, and Profit*, Chicago: The University of Chicago Press (1st edn. 1921).

Kobrak, Christophe (2012) 'USA: The International Attraction of the US Insurance Market', in Borscheid, Peter and Viggo Haueter, Niels (eds.) *World Insurance: The Evolution of a Global Risk Network*, Oxford: Oxford University Press, 274–308.

Kommerskollegium (2012) *Everybody is in Services – The Impact of Servicification in Manufacturing on Trade and Trade Policy*, Stockholm: National Board of Trade.

Koppell, Jonathan G. S. (2003) *The Politics of Quasi-Government: Hybrid Organizations and the Dynamics of Bureaucratic Control*, Cambridge: Cambridge University Press.

KPMG (2015) *Evolving Insurance Regulation: The Journey Begins*, April.

KPMG International (2013) *KPMG Survey of Corporate Responsibility Reporting 2013*, KPMG International.

Kraidy, Marwan (2005) *Hybridity, or the Cultural Logic of Globalization*, Philadelphia: Temple University Press.

Krause Hansen, Hans (2008) 'Investigating the Disaggregation, Innnovation, and Mediation of Authority in Global Politics', in Krause Hansen, Hans and Salskov-Iversen, Dorte (eds.) *Critical Perspectives on Private Authority an Global Politics*, Houndmills: Palgrave Macmillan, 1–23.

Krause Hansen, Hans and Salskov-Iversen, Dorte (eds.) (2008) *Critical Perspectives on Private Authority in Global Politics*, Houndmills: Palgrave Macmillan.

Krislov, Samuel (1997) *How Nations Choose Product Standards and Standards Change Nations*, Pittsburgh, PA: Pittsburgh University Press.

Kshetri, N. and Dholakia, N. (2009) 'Professional and trade associations in a nascent and formative sector of a developing economy: A case study of the NASSCOM effect on the Indian offshoring industry', *Journal of International Management*. 15(2): 225–239.

Lalonde, Carole and Boiral, Olivier (2012) 'Managing risks through ISO 31000: A critical analysis', *Risk Management* 14(4): 272–300.

Lanoszka, Alexander (2016) 'Russian hybrid warfare and extended deterrence in eastern Europe', *International Affairs* 92(1): 175–195.

Latour, Bruno (1984) *Les Microbes: guerre et paix*, Paris: La Découverte.

(1991) *Nous n'avons jamais été modernes: Essai d'anthropologie symétrique*, Paris: La Découverte.

(1993) *We Have Never Been Modern* (1st edn, 1991), London: Harvester Wheatsheaf.

(2016) '*Onus Orbis Terrarum*: About a possible shift in the definition of sovereignty', *Millennium* 44(3): 305–320.

(2017) *Où atterrir?: Comment s'orienter en politique*, Paris: La Découverte.

Leander, Anna (2014) 'Understanding US National Intelligence: analyzing practices to capture the chimera', in Best, Jacqueline and Gheciu, Alexandra (eds) *The Return of the Public in Global Governance*, Cambridge: Cambridge University Press: 197–220.

(2016) 'Seen and Unseen: Hybrid Rule in International Security', in Hurt, Shelley and Lipschutz, Ronnie (eds.) *Hybrid Rule and State Formation: Public-Private Power in the 21st Century*, London: Routledge: 143–159.

Lehtonen, Turo-Kimmo (2014) 'Picturing how life insurance matters', *Journal of Cultural Economy* 7(3): 308–333.

Lehtonen, Turo-Kimmo and Van Hoyweghen, Ine (2014) 'Editorial - insurance and the economization of uncertainty', *Journal of Cultural Economy* 7(4): 532–540.

Leimgruber, Matthieu (2013) 'The Embattled Standard-Bearer of Social Insurance and Its Challenger: The ILO, The OECD, and the "Crisis of the Welfare State", 1975–1985', in Kott, Sandrine and Droux, Joëlle (eds.) *Globalizing Social Rights. The International Labour Organization and Beyond*, Basingstoke: Palgrave Macmillan/ILO, 293–309.

Lengwiler, Martin (2009) 'Double Standards: The History of Standardizing Humans in Modern Life Insurance', in Lampland, Martha and Star, Susan Leigh (eds.) *Standards and their Stories*, Ithaca, NY: Cornell University Press, 95–113.

Levi-Faur, David (2011a) 'Regulation and Regulatory Governance', in Levi-Faur, David (ed.) *Handbook on the Politics of Regulation*, Cheltenham: Edward Elgar: 3–21.

(ed.) (2011b) *Handbook on the Politics of Regulation*, Cheltenham: Edward Elgar.

Levy, David L. (2005) 'Offshoring in the new global political economy', *Journal of Management Studies* 42(3): 685–693.

Lezon, Catherine (2015) 'Quelle régulation mondiale pour les assurances?', *Réalités Industrielles Février*: 54–57.
Lim, Alwyn and Tsutsui, Kiyoteru (2012) 'Globalization and commitment in corporate social responsibility: Cross-national analyses of institutional and political-economy effects', *American Sociological Review* 77(1): 69–98.
Lionnet, Françoise (1995) *Postcolonial Representations: Women, Literature, Identity*, Ithaca, NY: Cornell University Press.
Lobo-Guerrero, Luis (2011) *Insuring Security: Biopolitics, Security and Risk*, London: Routledge.
(2012) *Insuring War: Sovereignty, Security and Risk*, London: Routledge.
(2014) 'Life securitisation, the event object of insurance and the strategisation of time', *Journal of Cultural Economy* 7(3): 353–370.
(2016) *Insuring Life: Value, Security and Risk*, London: Routledge.
Loconto, Allison and Busch, Lawrence (2010) 'Standards, techno-economic networks, and playing fields: Performing the global market economy', *Review of International Political Economy* 17(3): 507–536.
Low, Sydney (1913) *The Governance of England*, New York: G.P. Putnam's Sons.
Lukes, Steven (2005) *Power: A Radical View*. 2nd edn (1st edn 1974), London: Palgrave Macmillan.
Lupton, Sylvie (2005) 'Shared quality uncertainty and the introduction of indeterminate goods', *Cambridge Journal of Economics* 29(3): 399–421.
Lysandrou, Photis (2016) 'The colonization of the future: An alternative view of financialization and its portents', *Journal of Post-Keynesian Economics* 39(4): 444–472.
Magnette, Paul (2017) *CETA: Quand l'Europe déraille*, Waterloo: Luc Pire.
Mahler, Imelda (2007) 'Economic governance: Hybridity, accountability, and control', *Columbia Journal of European Law* 13: 680–703.
Mallard, Grégoire (2014): *Nuclear Diplomacy in the Age of Global Fracture*, Chicago: Chicago University Press.
Mandin, Lou and Palier, Bruno (2009) 'Le rôle des organisations internationales dans les réformes des systèmes nationaux de protection sociale', in Schemeil, Yves and Eberwein, Wolf-Dieter (eds.) *Normer le monde*, Paris: L'Harmattan, 177–210.
Marks, Gary, Hooghe, Liesbet, and Blank, Kermit (1996) 'European integration from the 1980s: State-centric versus multi-level governance', *Journal of Common Market Studies* 34(3): 341–378.
Marschallek, Christian (2011) 'Back to the State? The Public Policies of Private and Public Pensions in Britain', in Leisering, Lutz (ed.) *The New Regulatory State. Regulating Pensions in Germany and the UK*, Basingstoke: Palgrave Macmillan, 103–126.
Mathis, James (2012) 'Multilateral aspects of advanced regulatory cooperation: Considerations for a Canada-EU Comprehensive Trade Agreement (CETA)', *Legal Issues of Economic Integration* 39(1): 73–91.
Mattli, Walter (2001) 'The politics and economics of international institutional standards setting: An introduction', *Journal of European Public Policy* 8(3): 328–344.

Mattli, Walter and Büthe, Tim (2003) 'Setting international standards: Technological rationality or primacy of power?', *World Politics* 56(1): 1–42.

(2011) *The New Global Rulers: The Privatization of Regulation in the World Economy*, Princeton, NJ: Princeton University Press.

Mattli, Walter and Woods, Ngaire (eds.) (2009) *The Politics of Global Regulation*, Princeton, NJ: Princeton University Press.

Mayer, Frederick W., Phillips, Nicola, and Posthuma, Anne C. (2017) 'The political economy of governance in a "global value chain world"', *New Political Economy* 22(2): 129–133.

Mazzaferro, Francesco and Brinkhoff, Jeroen (2012) 'LTG Symposium: A Macro-Prudential Perspective', Solvency II Wire, London, 14 September, www.solvencyiiwire.com/ltg-symposium-a-macro-prudential-perspective/5444514, accessed April 2015.

McIvor, Ronan (2010) *Global Services Outsourcing*, Cambridge: Cambridge University Press.

Millar, Carla C. J. M. and Choi, Chong Ju (2011) 'The innovative future of service industries: (Anti-)globalization and commensuration', *Service Industries Journal* 31(1): 21–38.

Miller, Peter, Kurunmaki, Liisa, and O'Leary, Ted (2010) 'Calculating Hybrids', in Higgins, Vaughan and Larner, Wendy (eds.) *Calculating the Social: Standards and the Reconfiguration of Governing*, Basingstoke: Macmillan: 21–37.

Mithas, Sunil and Whitaker, Jonathan (2007) 'Is the world flat or spiky? Information intensity, skills, and Global Service Disaggregation', *Information Systems Research* 18(3): 237–259.

Mock, Dave (2005) *The Qualcomm Equation: How a Fledgling Telecom Company Forged a New Path to Big Profits and Market Dominance*, New York: AMACOM, American Management Association.

Monkiewicz, Jan (2013) 'Dialectics of the current regulatory and supervisory developments in insurance', *The Geneva Papers on Risk and Insurance* 38: 183–188.

Montgomerie, Johnna (ed.) (2017) *Critical Methods in Political and Cultural Economy*, Abingdon: Routledge.

Morgan, Kevin (2004) 'The exaggerated death of geography: Learning, proximity and territorial innovation systems', *Journal of Economic Geography* 4: 3–21.

Mouhoud, El Mouhoub (ed.) (2010) *Economie des services et développement des territoires*, Paris: La documentation Française/Délégation interministérielle à l'aménagement du territoire et à l'attractivité régionale (DATAR) - Travaux n°11.

Moulier-Boutang, Yann (2007) *Le capitalisme cognitif: La nouvelle grande transformations*, Paris: Editions Amsterdam.

Murphy, Craig (2000) 'Global governance: Poorly done and poorly understood', *International Affairs* 76(4): 789–803.

Murphy, Craig and Yates, JoAnne (2009) *The International Organization for Standardization (ISO): Global Governance through Voluntary Consensus*, London: Routledge.

Nasscom (2010) *Strategic Review*, New Delhi: National Association of Software and Services Companies.

(2012) *The IT-BPM Sector in India: Strategic Review 2012*, New Delhi: National Association of Software and Services Companies.

(2018) *The IT-BPM Sector in India: Strategic Review 2018 – Amplify Digital*, New Delhi: National Association of Software and Services Companies.

National Institute of Standards and Technology (2017) *Twentieth Annual Report on Federal Agency Use of Voluntary Consensus Standards and Conformity Assessment*, Gaithersburg: National Institute of Standards and Technology, US Department of Commerce.

(n.d.) *NIST Three Year Programmatic Plan 2017–2019*, available online at: www.nist.gov/sites/default/files/documents/director/planning/3_year_plan_2017-19_web_ready2.pdf.

National Science and Technology Council (2011) *Federal Engagement in Standards Activities to Address National Priorities. Background and Proposed Policy Recommendations*, Washington DC: Subcommittee on Standards National Science and Technology Council, 10 October.

Nayyar, Gaurav (2012) *The Service Sector in India's Development*, Cambridge: Cambridge University Press.

Nederveen Pieterse, Jan (2001) 'Hybridity, so what?: The anti-hybridity backlash and the riddles of recognition', *Theory, Culture & Society* 18(2–3): 219–245.

Negandhi, Anant R. and Palia, Aspy P. (1988) 'Changing multinational corporation – nation state relationship: The case of IBM in India', *Asia Pacific Journal of Management* 6(1): 15–38.

Ní Mhurchú, Aoileann (2015) 'Ambiguous subjectivity, irregular citizenship: From inside/outside to being-caught in-between', *International Political Sociology* 9(2): 158–175.

Niang, Nafi (2013) *Normalisation et dimension culturelle des délocalisations de services: les cas du Maroc et de l'Inde*, Thèse de Doctorat, présentée à la Faculté des sciences sociales et politiques, pour l'obtention du grade de Docteur ès science politique, Université de Lausanne.

Nicolaïdis, Kalypso and Egan, Michelle (2001) 'Transnational market governance and regional policy externalities: Why recognize foreign standards?', *Journal of European Public Policy* 8(3): 454–473.

Nicolaidis, Kalypso and Shaffer, Gregory (2005) 'Transnational mutual recognition regimes: Governance without global government', *Law and Contemporary Problems* 68(3/4): 263–317.

OECD (2000) *Information Technology Outlook. ICT, E-Commerce and the Information Economy*, Paris: Organisation for Economic Co-operation and Development.

(2010) *Measuring Globalisation: OECD Economic Globalisation Indicators*, Paris: Organisation for Economic Co-operation and Development.

(2012) *Recommendation of the Council on Regulatory Policy and Governance*, Paris: Organisation for Economic Co-operation and Development.

(2014a) *Mortality Assumptions and Longevity Risk: Implications for Pension Funds and Anuity Providers*, Paris: Organisation for Economic Co-operation and Development.

(2014b) *OECD Pensions Outlook 2014*, Paris: Organisation for Economic Co-operation and Development.

(2017) *OECD Services Trade Restrictiveness Index*: Organisation for Economic Co-operation and Development, June, available online at: www.oecd.org/tad/policynotes/STRI trade policy note.pdf.

OECD-WTO (2013) *Trade in Value Added: Concepts, Methodologies and Challenges (Joint OECD-WTO note)*.

Orenstein, Mitchell A. (2008) *Privatizing Pensions: The Transnational Campaign for Social Security Reform*, Princeton, NJ: Princeton University Press.

Orléan, André (2011) *L'empire de la valeur: Refonder l'économie*, Paris: Seuil.

Oshri, Ilan, Kotlarsky, Julia, and Willcocks, Leslie (2015) *The Handbook of Global Outsourcing and Offshoring*, Basingstoke: Palgrave Macmillan.

Osiander, Andreas (2001) 'Sovereignty, international relations, and the Westphalian myth', *International Organization* 55(2): 251–287.

Osmont, Annick (1998) 'La « governance »: concept mou, politique ferme', *Annales de la Recherche Urbaine* 80–81: 19–26.

Palan, Ronen (2003) *The Offshore World: Sovereign Markets, Virtual Places, and Nomad Millionaires*, Ithaca, NY: Cornell University Press.

Parasuraman, A, Zeithaml, Valerie A., and Berry, Leonard L. (1985) 'A conceptual model of service quality and its implication for future research', *Journal of Marketing* 49(4): 41–50.

Parthasarathy, Balaji (2004) 'India's Silicon Valley or Silicon Valley's India? Socially embedding the computer software industry in Bangalore', *International Journal of Urban and Regional Research* 28(3): 664–685.

(2013a) 'A Half Century of Envisioning the Future in Bangalore: From Autarky to Innovation in the Global Information Economy', in Cooke, Philip, Searle, Glen, and O'Connor, Kevin (eds.) *The Economic Geography of the IT Industry in the Asia Pacific Region*, London: Routledge, 221–242.

(2013b) 'The Changing Character of Indian Offshore ICT Services Provision, 1985–2010', in Bardhan, Ashok, Jaffee, Dwight M., and Kroll, Cynthia A. (eds.) *The Oxford Handbook of Offshoring and Global Employment*, Oxford: Oxford University Press, 380–404.

Parthasarathy, Balaji and Srinivasan, Janaki (2008) 'Contestation in the shaping of standards for the information age', *Science, Technology & Society* 13(2): 279–301.

Paterson, Matthew (2001) 'Risky business: Insurance companies in global warming politics', *Global Environmental Politics* 1(4): 8–42.

Patton, Michael Quinn (2002) *Qualitative Research and Evaluation Methods*. 3rd edn, Thousand Oaks, CA: Sage.

Paulk, Mark C. (2001) 'A History of the Capability Maturity Model for Software', available online at: www.sel.unsl.edu.ar/ApuntesMaes/Anteriores/MetodologiasAgiles/History of CMM.pdf, accessed 20 June 2016.

Pauwelyn, Joost (2014) 'Rule-based trade 2.0? The rise of informal rules and international standards and how they may outcompete WTO treaties', *Journal of International Economic Law* 17(4): 739–751.

Paye, Olivier (2005) 'La gouvernance. D'une notion polysémique à un concept politologique', *Etudes Internationales* xxxvi(1): 13–40.

Payne, Anthony (2005) 'The Study of Governance in a Global Political Economy', in Phillips, Nicola (ed.) *Globalizing International Political Economy*, Basingstoke: Palgrave Macmillan, 55–81.

Payne, Anthony and Phillips, Nicola (eds.) (2014) *Handbook of the International Political Economy of Governance*, Aldershot: Edward Elgar.

Pérez, Roland (2003) *La gouvernance de l'entreprise*, Repères, Paris: La Découverte.

Petit, Pascal (2007) 'The Political Economy of Services', in Bryson, John R. and Daniel, Peter W. (eds.) *The Handbook of Service industries*, Cheltenham: Edward Elgar, 77–97.

(2008) 'Les services et la transformation contemporaine des formes de concurrence', in Du Tertre, Christian and Laurent, Catherine (eds.) *Secteurs et territoires dans les régulations émergentes*, Paris: L'Harmattan, 217–236.

(2013) 'Le capital risque à la recherche d'un second soufle', in Graz, Jean-Christophe and Niang, Nafi (eds.) *Services sans frontières*, Paris: Presses de Science Po, 253–280.

Pijl, Kees van der (2014) *The Discipline of Western Supremacy: Modes of Foreign Relations and Political Economy*, vol. 3, London: Pluto Press.

Pitschas, Christian (2016) 'Transatlantic Trade and Investment Partnership (TTIP): The devil in disguise or a golden opportunity to build a transatlantic marketplace?', *British Journal of American Legal Studies* 5(2): 315–340.

Ponte, Stefano, Gibbon, Peter, and Vestergaard, Jakob (eds.) (2011) *Governing through Standards: Origins, Drivers and Limitations*, Houndmills: Palgrave Macmillan.

Porter, Tony (2005) 'Private authority, technical authority, and the globalisation of accounting standards', *Business and Politics* 7(3): article 2.

Potoski, Matthew and Prakash, Aseem (eds.) (2009) *Voluntary Programs: A Club Theory Perspective*, Cambridge, MA: MIT Press.

Prabhu, Anjali (2007) *Hybridity: Limits, Transformations, Prospects*, Albany: State University of New York Press.

Prakash, Aseem and Potoski, Matthew (2006) *The Voluntary Environmentalists: Green Clubs, ISO 14001, and Voluntary Environmental Regulations*.

(2010) 'The International Standards Organization as a Global Governor: A Club Theory Approach', in Avant, Deborah D., Finnemore, Martha, and Sell, Susan K. (eds.) *Who Governs the Globe?*, Cambridge: Cambridge University Press, 72–101.

Ramadorai, Subramaniam (2011) *The TCS Story and Beyond*, New Delhi: Penguin.

Ramella, Marcelo (2010) 'Securitization and Insurance', in Kempler, Cecelia, et al. (eds.) *Global Perspectives on Insurance Today: A Look at National Interest versus Globalization*, Houndmills: Palgrave Macmillan, 229–244.

Ramirez, Carlos (2013) 'Normalisation des services marchands ou marchandisation des normes ? Les Big 4 et la normalisation internationale de la comptabilité et de l'audit', in Graz, Jean-Christophe and Niang, Nafi (eds.) *Services sans frontières*, Paris: Presses de Sciences Po.

Ravenhill, John (2016) 'The political economy of an "Asian" Mega-FTA: The Regional Comprehensive Economic Partnership', *Asian Survey* 56(6): 1077–1100.

Risse, Thomas, Börzel, Tanja A., and Draude, Anke (eds.) (2018) *The Oxford Handbook of Governance and Limited Statehood*, New York: Oxford University Press.

Rubalcaba, Luis (2007) *The New Service Economy: Challenges and policy Implications for Europe*, London: Edward Elgar.
Rusconi, Rob (2008) *National Annuity Markets: Features and Implications*, OECD Working Papers on Insurance and Private Pensions, Working Paper No. 24.
Ruwet, Coline (2009) *Des filetages à la RSE. Normalisation et démocratie. Sociologie du processus d'élaboration d'ISO 26000*, Thèse de Doctorat, Département de sociologie, Université catholique de Louvain.
Said, Edward W. (1978) *Orientalism*, London: Routledge.
Salomon, Jean-Jacques (1992) *Le destin technologique*, Paris: Balland.
Saraswati, Jyoti (2008) 'The Indian IT industry and neoliberalism: The irony of a mythology', *Third World Quarterly* 29(6): 1139–1152.
Sarma, Mahesh and Krishna, Venni Venkata (2010) 'State and the software: Public policies in the shaping of the Indian software sector', *The Service Industries Journal* 30: 25–42.
Sassen, Saskia (2003a) 'Globalization or denationalization?', *Review of International Political Economy* 10(1): 1–22.
(2003b) 'The participation of states and citizens in global governance', *Indiana Journal of Global Legal Studies* 10(5): 5–28.
Schepel, Harm (2005) *The Constitution of Private Governance: Product Standards in the Regulation of Integrating Markets*, Oxford: Hart Publishing.
Schirm, Stefan A. (ed.) (2004) *New Rules for Global Markets: Public and Private Governance in the World Economy*, London: Palgrave Macmillan.
Schmidt, Susanne K. and Werle, Raymund (1998) *Coordination Technology: Studies in the International Standardization of Telecommunication*, Cambridge, MA: MIT Press.
Schoechle, Timothy (2009) *Standardization and Digital Enclosure: The Privatization of Standards, Knowledge, and Policy in the Ages of Global Information Technology.*, Hershey, NY: Information Science Reference.
Senarclens, Pierre de (1999) *Mondialisation, souveraineté et théories des relations internationales*, Paris: A. Colin.
Sharma, Dinesh C. (2015) *The Outsourcer: The Story of India's IT Revolution*, Cambridge, MA: MIT Press.
Silburn, P. A. (1910) *The Governance of Empire*, London: Longmans, Green & Co.
Smolinski, Luke (2013) 'Europe divided over Solvency II saga', *IQ – Insider Quarterly*, Autumn, www.insiderquarterly.com/europe-divided-over-solvency-ii-saga, accessed 25 June 2015.
Software Engineering Institute (2010) *CMMI® for Services, Version 1.3*, n.d.: Carnegie Mellon University, November.
Spruyt, Hendrik (2001) 'The supply and demand of governance in standard-setting: Insights from the past', *Journal of European Public Policy* 8(3): 371–391.
Srinivas, Smita (2012) *Market Menagerie: Health and Development in Late Industrial States*, Stanford, CA: Stanford University Press.
Stiglitz, Joseph (1987) 'The causes and consequences of the dependence of quality on price', *Journal of Economic Literature* 25: 1–48.
Strange, Susan (1996) *The Retreat of the State: The Diffusion of Power in the World Economy*, Cambridge: Cambridge University Press.

Stuart-Linton, Charles E.T. (1912) *The Problem of Empire Governance*, London: Longmans, Green & Co.

Sundbo, J. (2002) 'The service economy: Standardisation or customisation?', *The Service Industries Journal* 22(4): 93–116.

Sunstein, Cass R. and Thaler, Richard (2008) *Nudge: Improving Decisions About Health, Wealth, and Happiness*, New Haven, CT: Yale University Press.

Swann, G.M. Peter (2000) *The Economics of Standardization. Final Report for Standards and Technical Regulations Directorate Department of Trade and Industry*, Manchester: Manchester Business School, 11 December.

(2010) *The Economics of Standardization: An Update*, Report for the UK Department of Business, Innovation and Skills (BIS).

Swiss Re (2003) *Natural Catastrophes and Reinsurance*, Zurich: Swiss Reinsurance company.

(2011) 'State involvement in insurance markets', *Sigma*(3).

(2012a) 'Facing the interest rate challenge', *Sigma*(4).

(2012b) *Lessons from Recent Major Earthquakes*: Swiss Reinsurance Company, January.

(2015) 'Natural catastrophe and man-made disasters in 2014', *Sigma*(2).

(2018a) 'Natural catastrophe and man-made disasters in 2017: a year of record-breaking losses', *Sigma* (1).

(2018b) 'World insurance in 2017: Solid, but mature life markets weigh on growth', *Sigma* (3).

Tamm Hallström, Kristina (2004) *Organizing International Standardization: ISO and the IASC in Quest of Authority*, Cheltenham: Edward Elgar.

Tate, Jeffrey (2001) 'National Varieties of Standardization', in Hall, Peter and Soskice, David (eds.) *Varieties of Capitalism. The Institutional Foundations of Comparative Advantage*, Oxford: Oxford University Press, 442–473.

Taylor, Phil and Bain, Peter (1999) '"An assembly line in the head": Work and employment relations in the call centre', *Industrial Relations Journal* 30(2): 101–117.

Teschke, Benno (2002) 'Theorising the Westphalian system of states: International relations from absolutism to capitalism', *European Journal of International Relations* 8(1): 5–48.

Tether, Bruce S., Hipp, Christiane, and Miles, Ian (2001) 'Standardisation and particularisation in services: Evidence from Germany', *Research Policy* 30(7): 1115–1138.

The Economist (1985) 'A GATT for services', 12 October.

(2013) 'Special Report: Outsourcing and Offshoring', 19 January.

Tilley, Aaron (2016) 'Microsoft, Qualcomm and Intel Start Playing Nice On "Internet Of Things" Standards', *Forbes*, 19 February.

Timmermans, Stefan and Epstein, Steven (2010) 'A world of standards but not a standard world: Toward a sociology of standards and standardization', *Annual Review of Sociology* 36: 69–89.

Tömmel, Ingeborg and Verdun, Amy (eds.) (2009) *Innovative Governance in the European Union: The Politics of Multilevel Policymaking*, Boulder, CO: Lynne Rienner.

Trachtman, Joel P. (2003) 'Toward open recognition? Standardization and regional integration under Article XXIV of GATT', *Journal of International Economic Law* 6(2): 459–492.

Trubek, David M. and Trubek, Louise G. (2007) 'New governance and legal regulation: Complementarity, rivalry or transformation', *Columbia Journal of European Law* 13: 539–564.

Tsutsui, Kiyoteru and Lim, Alwyn (eds.) (2015) *Corporate Social Responsibility in a Globalizing World*, Cambridge: Cambridge University Press.

UNCTAD (2004) *World Investment Report: The Shift towards Services*, Geneva: United Nations.

United Nations (2010) *Manual on Statistics of International Trade in Services.*

United States Government Accountability Office (2018) *National Institute of Standards and Technology. Additional Review and Coordination Could Help Meet Measurement Service Needs and Strengthen Standards Activities. Report to the Subcommittee on Research and Technology, Committee on Science, Space, and Technology, House of Representatives*, Washington, DC: GAO-18–445.

Upadhya, Carol (2009) 'Imagining India: Software and the ideology of liberalisation', *South African Review of Sociology* 40(1): 76–93.

Upadhya, Carol and Vasavi, A. R. (eds.) (2008) *In an Outpost of the Global Economy: Work and Workers in India's Information Technology Industry*, New Delhi: Routledge India.

Uranie (1996) 'Numéro spécial "Hybrides et Hybridités"', *Uranie* 6: 1-240.

Utting, Peter (2008) 'Rearticulating Regulatory Approaches: Private-Public Authority and Corporate Social Responsibility', in Rittberger, Volcker and Nettesheim, Martin (eds.) *Authority in the Global Political Economy*, Houndmills: Palgrave Macmillan, 241–275.

VanDuzer, J. Anthony (2012) 'A critical look at the prospects for robust rules for services in preferential trading agreements', *Legal Issues of Economic Integration* 39(1): 29–49.

Veblen, Thorstein (1919) *The Place of Science in Modern Civilization and other Essays*, New York: B. W. Huebsch.

Viardot, Eric, Sherif, Mostafa Hashem, and Chen, Jin (2016) 'Managing innovation with standardization: An introduction to recent trends and new challenges', *Technovation* 48–49: 1–3.

Villar, Constanze (2005) 'Pour une théorie du discours diplomatique', *Annuaire Français de Relations Internationales* 6: 45–61.

Vion, Antoine et al. (2013) 'Normalisation informatique et marchés de services: le cas du XML', in Graz , Jean-Christophe and Niang, Nafi (eds.) *Services sans frontières: Mondialisation, normalisation et régulation de l'économie des services*, Paris: Presses de Science Po, 281–316.

Vogel, David (1995) 'National Regulation in the Global Economy', in Vogel, David (ed.) *Trading Up: Consumer and Environmental Regulation in a Global Economy*, Cambridge, MA: Harvard University Press, 1–23.

(2009) 'The Private Regulation of Global Corporate Conduct ', in Mattli, Walter and Woods, Ngaire (eds.) *The Politics of Global Regulation*, Princeton, NJ: Princeton University Press, 152–88.

Vries, Henk de (1999) *Standardization: A Business Approach to the Role of National Standardization Organizations*, Boston: Kluwer.

Vries, Henk de and Wiegmann, Paul Moritz (2017) 'Impact of Service Standardization on Service Innovation', in Hawkins, Richard, Blind, Knut, and Page, Robert (eds.) *Handbook of Innovation and Standards*, Aldershot: Edwrd Elgar, 187–211.

Weiss, Linda (2014) *America Inc.?: Innovation and Enterprise in the National Security State*, Ithaca, NY: Cornell University Press.

Weiss, Thomas G. (2011) *Thinking about Global Governance: Why People and Ideas Matter*, London: Routledge.

Werle, Raymund (2001) 'Institutional aspects of standardization – jurisdictional conflicts and the choice of standardization organizations', *Journal of European Public Policy* 8(3): 392–410.

Winickoff, David E. and Mondou, Matthieu (2016) 'The problem of epistemic jurisdiction in global governance: The case of sustainability standards for biofuels', *Social Studies of Science* 47(1): 7–32.

Winn, Jane K (2009) 'Globalization and standards: The logic of two-level games', *I/S: A Journal of Law and Policy for the Information Society* 5(2).

World Bank (1989) *From Crisis to Sustainable Growth - Sub Saharan Africa: A Long-Term Perspective Study*, Washington, DC: The World Bank.

(1997) *World Development Report 1997: The State in a Changing World*, Washington, DC: The World Bank.

(2002) *World Development Report 2002: Building Institutions for Markets*, Washington, DC: The World Bank.

World Trade Organization (2012) *World Trade Report 2012: Trade and Public Policies: A Closer Look at Non-Tariff Measures in the 21st Century*, Geneva: WTO.

Xiang, Biao (2007) *Global 'Body Shopping': An Indian Labor System in the Information Technology Industry*, Princeton, NJ: Princeton University Press.

Yates, JoAnne and Murphy, Craig (2019) *Engineering Rules: Global Standard Setting since 1880*, Baltimore, MD: Johns Hopkins University Press.

Young, Robert (1995) *Colonial Desire: Hybridity in Theory, Culture, and Race*, London: Routledge.

Zeithaml, V. A., Parasuraman, A., and Berry, L. L. (1990) *Delivering Service Quality*, New York: The Free Press.

Zhang, Vicki (2014) *Uncalculated Risks: The Transformation of insurance, the Erosion of Regulation, and the Economic and Social Consequences*, Toronto: Canadian Scholars' Press.

Zinecker, Heidrun (2009) 'Regime-hybridity in developing countries: Achievements and limitations of new research on transition', *International Studies Review* 11: 302–331.

Zuckerman, Amy (1996) *International Standards Desk Reference: Your Passport to World Markets: Iso 9000, Ce Mark, Qs-9000, Ssm, Iso 14000, Q 9000, American, European and Global Standards Systems*, New York: American Management Association.

Zumbansen, Peer (2011) 'Neither 'public' nor 'private', 'national' nor 'international': Transnational corporate governance from a legal pluralist perspective', *Journal of Law and Society* 38(1): 50–75.

Index

Acheraïou, Amar, 6, 34, 40
ACORD. *See* Association for Cooperative Operations Research and Development
Afnor
  distinctiveness of standards and, 97–100
  sectorial specificity and, 97–98
AIG, 122, 168–208
Akerlof, George, 63
Allen, Woody, 114
ambiguity. *See also* transnational hybrid authority
  Best on, 53
  governance and, 19, 26, 32–33
  hybrid power based on, 2–3, 25–26, 36, 52–53, 113, 210, 213–217
  myths and, 36
  as substantive attribute, 19, 25, 34–35, 52, 210
ambiguity of standards, 3, 16, 81, 210–211, 217–218
  internationalisation of standards and, 12
  self-regulation and, 10, 26–28, 217
  social contestability and, 216–217
American National Standards Institute (ANSI), 100–103, 110, 209
  NAFTA and, 218
American Society for Testing and Materials. *See* ASTM International
American Society of Mechanical Engineers (ASME), 101–103
Amin, Ash, 215
annuities, 150
  life insurance and, 124, 144
  UK and, 144
ANSI. *See* American National Standards Institute
Apple, 61
artificial intelligence, 185, 220
ASEAN. *See* Association of Southeast Asian Nations
ASME. *See* American Society of Mechanical Engineers
Association for Cooperative Operations Research and Development (ACORD), 153–154, 159–160, 171
Association of Southeast Asian Nations (ASEAN), 103–104
ASTM International, 102–103
authority. *See also* private international authority; transnational hybrid authority
  diffusion of, 44
  power versus, 4
  technical, 47
  transnational private, 4–6, 42

Bakhtin, Mikhail, 39, 213
Banerjee, Sudeep, 184
Bank for International Settlement (BIS), 132–133
Barry, Dean, 120
Bartley, Tim, 24, 217
Basel Committee on Banking Supervision, 132–133
Beck, Ulrich, 13, 36–37
Bell, Daniel, 60
Benari, Gideon, 122, 127
Berle, Adolf, 30
Bermuda, sustainable reporting standards and, 168–169
Best, Jacqueline, 6, 53, 214, 216
Beverley, Jane, 129
Bhabha, Homi K., 39
Bhathnagar, Raju, 203
big data, 220–221
BIS. *See* Bank for International Settlement; Bureau of Indian Standards
Blake, David, 123–124, 147, 150
Blind, Knut, 7, 73–74
Blinder, Alan, 54–55
Boden, Mark, 61

body-shopping, India, service offshoring in, and, 179–180, 182–183
Bolkestein Directive (Directive 2006/123/EC), 95–96
boundary objects, standards as, 219
BPM. *See* business process management
BPO. *See* business process outsourcing
Brahmankar, Manoj, 197
Bratton, Benjamin H., 221
Brexit, 103–104
Brinkhoff, Jeroen, 130
British Empire, governance and, 30
British Standard Institution (BSI), 96, 186, 212
Bryson, John R., 60–61, 72
BSI. *See* British Standard Institution
Bureau of Indian Standards (BIS), 195–197, 203, 208
Busch, Lawrence, 9, 80–81, 83, 219, 221–222
business process management (BPM), 175
business process outsourcing (BPO), 15, 173–174, 213
  India and, 22–23, 183–184, 188–191, 196–198, 202–206
business services, consumer services versus, 17
Büthe, Tim, 45, 91

call centres, 17–18, 27–28, 40, 60
  India in, 188–191
Callon, Michel, 38, 64–65
Canada-European Union Comprehensive Economic and Trade Agreement (CETA), 3, 104–105, 211
  mutual recognition of standards and, 108
  regulatory chill effect and, 106
  regulatory convergence and, 48
  regulatory cooperation and, 106–107
Capability Maturity Model Integration in the field of services (CMMI for Services), 6–7, 190
Capability Maturity Model Integration Institute (CMMI Institute), 187–190, 194–195
capability maturity models, India, service offshoring in, and, 18, 187–190
capital requirements, insurance regulation and supervision in
  BCR as, 133
  extrapolation in, 131
  IAIS and, 132–135
  internal models for, 136, 139
  matching adjustments and, 128–130
  MCR as, 126
  SCR as, 124–128
  in Solvency II, 125–132
  US and, 135
  volatility adjustments in, 122–131
Castells, Manuel, 58
Catastrophe Risk Evaluating and Standardizing Target Accumulation (Cresta), 156–158, 171
CEN. *See* Comité européen de normalisation
CEN Horizontal European Service Standardization Strategy (CHESSS), 96–97
CENELEC. *See* Comité Européen de Normalisation Electrotechnique
CETA. *See* Canada-European Union Comprehensive Economic and Trade Agreement
Chamberlin, Edward Hastings, 64–65
Chandler, David, 221
Chanteau, Jean-Pierre, 64
CHESS. *See* CEN Horizontal European Service Standardization Strategy
chimeras, 35
China, 195
Chowdhury, J Roy, 196
civil society, 42–44, 217
  INTERNORM project and, 19, 33–47
  preferential trade agreements and, 106
  standardisation and, 3, 99, 110, 217
  US and, 101
cloud computing, service offshoring and, 62
CMMI for Services. *See* Capability Maturity Model Integration in the field of services
CMMI Institute. *See* Capability Maturity Model Integration Institute
COBIT. *See* Control objectives for information and related technology
Codex Alimentarius, 106
codification, service offshoring and, 74
Colàs, Alejandro, 44
colonialism, hybridity and, 39–40
Comité européen de normalisation (CEN), 90–100
  CHESSS and, 96–97
  formal standards and, 7, 51, 90
  Regulation on European Standardisation (1025/2012) and, 96, 99, 110
Comité Européen de Normalisation Electrotechnique (CENELEC), 90–100
Commission on Global Governance, 31
commodification of standards, 218
Common Framework (ComFrame), in IAIS, 133

Commons, John, 9, 64
Comprehensive and Progressive Trans-Pacific Partnership (CPTPP), 4, 104
  regulatory cooperation and, 106
consent, transnational hybrid authority and, 45, 216
consortium standards, 7, 84–85, 133–134
Consumer Policy Committee (COPOLCO), 92
consumer protection
  insurance and, 131, 137–138
  mutual recognition and, 106–108
  standards and, 11–12, 46–47, 56, 64–65, 88–89, 91, 194–196
consumer services, business services versus, 17
Control objectives for information and related technology (COBIT), 190–191
COPOLCO. *See* Consumer Policy Committee
corporate governance, 30–31
  management and, 30
corporate social responsibility, 162–164
  AIG and, 168–208
  compariing EU and US, 163–164
Coughlan, Guy, 147–148
Cox, Robert W., 9, 40, 52, 214
CPTPP. *See* Comprehensive and Progressive Trans-Pacific Partnership
Cresta. *See* Catastrophe Risk Evaluating and Standardizing Target Accumulation
CSP standard. *See* Customer Service Provider standard
customer centres. *See* call centres
Customer Experience standard (CX standard), 190
Customer Service Provider standard (CSP standard), 190
Cutler, A. Claire, 42–43, 45
CX standard. *See* Customer Experience standard

Daniels, Peter W., 60–61, 72
data exchange standards, 160, 170–171
  ACORD and, 153–154, 159–160, 171
  Cresta and, 156–158, 171
  eEG7 and, 153–154
  GTDI as, 153
  institutional continuum and, 160
  material continuum and, 160–161, 211
  natural catastrophe insurance and, 156–162, 171
  spatial continuum and, 161
  SWIFT as, 153
  UN/CEFACT as, 153–154
  UN/EDIFACT as, 153
  UNICEDE and, 158
  UNTDID as, 153
  US and, 161
  W3C and, 152–153
  XML as, 152
Data Security Council of India (DSCI), 200
defence industry, India, service offshoring in, and, 178
denationalization, 4
depolitisation, hybrids and, 5, 26, 28–29
developed countries
  FDI and, 67–69
  GRI and, 165
  75/25 puzzle and, 66–67, 70
developing countries
  FDI and, 67–69
  GRI and, 165
  75/25 puzzle and, 66–67, 70
development, good governance and, 31
Dezalay, Yves, 33–34
Dholakia, N., 200
diffusion of authority, 28, 44, 213
digitability, service offshoring and, 71
digital, India, service offshoring in, 184–185
Dingwerth, Klaus, 165
direct participation
  EU and, 112
  US and, 89, 110, 112, 193–194
distinctiveness of service standards, 111. *See also* sectorial specificity
  Afnor and, 97–100
  EU and, 96–97
  ISO and, 93–94
Djelic, Marie-Laure, 33–34, 45
domestic services, 75/25 puzzle and, 69–70
Dossani, Rafiq, 177, 183
Doyle, Aaron, 114, 120
DSCI. *See* Data Security Council of India
du Tertre, Christian, 75–76, 161, 219–220
Dudouet, François-Xavier, 79–80

Economic Commission for Europe (ECE), 153
education, service offshoring and, 71–72, 181–182
eEG7, 153–154
Egan, Michelle, 24
Egyedi, Tineke, 91–100
Eicher, Lawrence D., 92
Eichinger, Margot, 165

EIOPA. *See* European Insurance and Occupational Pensions Authority
energy standards, 111–112
Epstein, Steven, 9
Ericson, Richard, 114, 120
eSourcing Capability Model (eSCM), India, service offshoring in, and, 197–198, 204
ETSI. *See* European Telecommunications Standards Institute
EU. *See* European Union
European Court of Justice, 95
European Directive Solvency II (Solvency II), 14, 22, 117, 122–123, 125–132, 138–139. *See also* Omnibus II Directive
  capital requirements in, 126–132
  Delegated Regulation (EU) 2015/35 and, 125, 131–132
  extrapolation in, 131
  IAIS and, 22, 132–135
  ICAS versus, 122–123
  ITS in, 125–126, 131–132
  life insurance and, 128–129
  long term guarantee measures, 127–132
  matching adjustments in, 128–130
  MCR in, 126
  ORSA and, 126
  pension funds and, 129, 137
  QIS5 for, 127, 131
  SCR in, 124–128
  Solvency II Framework Directive in, 125
  three pillars of, 126
  volatility adjustments in, 122–131
European Insurance and Occupational Pensions Authority (EIOPA), 125–126, 129, 220
  ITS of, 125–126, 131–132
  QIS5 published by, 127, 131
European Telecommunications Standards Institute (ETSI), 90
European Union (EU), 21. *See also* Comité européen de normalisation; European Directive Solvency II; European Insurance and Occupational Pensions Authority; Omnibus II Directive
  Bolkestein Directive and, 95–96
  corporate social responsibility and, 163–164
  direct participation and, 112
  energy standards and, 111–112
  EU Mandate M/340 and, 96
  EU Mandate M/371 and, 96
  food safety and, 104
  as institutional environment, 86–87, 89–91, 94–99, 110–113
  Lisbon Agenda and, 95
  multi-level governance and, 32
  mutual recognition of standards and, 85
  national participation and, 90, 110, 112
  New Approach and, 21, 90, 94–95, 99
  quality in, 96
  self-regulation and, 95
  service standards, distinctiveness of, and, 96–97
  Treaty on Stability, Coordination and Governance and, 32
evolutionary political economy, 177
exclusion principle, ISO And, 93, 96
Extensible Markup Language (XML), 152
extensive hypothesis, 12, 16, 212
  on service offshoring, 56–58, 78–79, 85, 173, 205–206, 211–212

FDI. *See* foreign direct investments
Federal Insurance Office, US (FIO), 134
financialisation, and insurance, 13–14, 21, 118–121
FIO. *See* Federal Insurance Office, US
Fisher, Paul, 122–123
food safety, EU and, 104
foreign direct investments (FDI) and internationalization of services, 55, 67
  developing versus developed countries and, 67–69
Foreign Exchange Regulation Act, in India, 178–179
formal standards, 7, 51
Foucault, Michel, 214–215
Foxconn, 61
Friedman, Thomas L., 15, 176, 182
Fukushima Daiichi nuclear disaster, 151
functional differentiation, 216

G20. *See* Group of Twenty
GAAP. *See* US Generally Accepted Accounting Principles
Gadrey, Jean, 58, 60, 75
Gandhi, Rajiv, 180
Garth, Bryant, 33–34
GATS. *See* General Agreement on Trade in Services
Gautier, Pascal, 98
gender equality, service industrialisation and, 75, 78
General Agreement on Trade in Services (GATS), 21, 59, 88–89
  MFN and, 108–109
  preferential trade agreements and, 108–109

Giegold, Sven, 132
global governance, 28–29, 31–34, 40–41, 214
Global Reporting Initiative (GRI), 84–85, 152, 162, 164–169
  developing versus developed countries and, 165
  market creation and, 141
global systemically important insurers (G-SIIs), 133
globalisation, 1, 104, 175–176, 214–215. *See also specific topics*
  governmental failures in, 1–2
Godin, Christian, 36
good governance, development and, 31
governance, 25
  ambiguity of, 19, 26, 32–33
  British Empire and, 30
  corporate, 30–31
  global, 28–29, 31–34, 40–41, 214
  hybrids and, 28–34
  insurance and, 120–122
  multi-level, 32
  power, as mode of exerting, 29
  sovereignty and, 29
  syncretic modes of, 6, 40
  Treaty on Stability, Coordination and Governance and, 32
Gramsci, Antonio, 216
Grande, Edgar, 45
GRI. *See* Global Reporting Initiative
Group of Twenty (G20), 209
G-SIIs. *See* global systemically important insurers
Guidelines for Trade Data Interchange (GTDI), 153
Gutner, Tamar, 31
Guzzini, Stefano, 29, 214

Hatzopoulas, Vassilis, 108
Hauert, Christophe, 9–10
Haufler, Virginia, 14, 120–121
Hausmann, Peter, 161
hegemony, 216
Heimer, Carol A., 120
Held, David, 28–29
Helleiner, Gerald K., 33
Hewitt de Alcántara, Cynthia, 29
Hibou, Béatrice, 5
Higgott, Richard, 4–5
Hill, Peter, 59–60
horizontal service standards, 95–96, 99, 111, 212
Howard, Philip N., 221
Hrubala, Bernard, 102
Hurt, Shelley, 5, 28
hybrid actors
  CMMI Institute as, 194–195
  institutional continuum spanned by, 44–45
  transnational hybrid authority and, 43–46
hybrid fora, 38
  qualification and, 65–66
hybrid objects
  material continuum spanned by, 46–47, 52
  transnational hybrid authority and, 46–48
hybrid power, 215–216. *See also* ambiguity, hybrid power based on; transnational hybrid authority
hybrid spaces
  endogenous versus exogenous logics of, 49–51
  spatial continuum spanned by, 49–52
  transnational hybrid authority and, 48–51
hybridity
  Bakhtin and, 39, 213
  Bhabha and, 39
  chimeras as, 35
  colonialism and, 39–40
  as default attribute, 5, 25–26, 34, 52
  Latour and, 6, 36–38
  modernity, crisis of, and, 6, 36–37
  regime-, 28
  regulation and, 26, 32
  as substantive attribute, 25, 34, 210
  territorial sovereignty and, 41
hybrids, 5–6, 19, 24–43
  depolitisation and, 28–29
  governance and, 28–34
  innovation, 27–28
  ISO and, 5
  myths and, 5–6, 35–36
  orthography of, 35
  transnational hybrid authorities as, 53

IAIGs. *See* internationally active insurance groups
IAIS. *See* International Association of Insurance Supervision
IBM (International Business Machines Corporation), 178–179
ICAS. *See* Individual Capital Adequacy Standards
ICPs. *See* insurance core principles
ICS. *See* insurance capital standard
ICTs. *See* information and communication technologies
immateriality of services, 17

Implementing Technical Standards (ITS), of EIOPA, 125–126, 131–132
India. *See also* National Association of Software and Services Companies (Nasscom)
  BPO and, 22–23, 183–184, 188–191, 196–198, 202–206
  DSCI in, 200
  Foreign Exchange Regulation Act in, 178–179
  IT Act in, 199
  National Skill Registry in, 199–201
India, service offshoring in, 15–16, 55, 77–78, 173–208, 213
  BIS and, 195–197, 203, 208
  body-shopping and, 179–180, 182–183
  capability maturity models and, 187–190
  CMMI Institute and, 187–190, 194–195
  COBIT and, 190–191
  Computer Software Export, Software Development and Training Policy and, 181
  CSP standard and, 190
  CX standard and, 190
  defence industry and, 178
  digital and, 184–185
  education and, 181–182
  eSCM and, 197–198
  IBM and, 178–179
  institutional continuum and, 177, 192–196, 205, 207
  ISO standards and, 193–194
  ISO/IEC 30105 and, 202–208, 220
  IT hardware industry and, 180
  land access, Nasscom and, 200
  liberalisation reforms and, 177, 181–182, 198
  material continuum and, 185–192, 205, 207
  origins of, 178–180
  process standards and, 185–192
  quality and, 182–191
  Six Sigma method and, 191
  spatial continuum and, 196–198, 205–208
  STPs and, 181–182, 200
  TCS and, 179
  trade unions and, 200–201, 207
  transnational hybrid authority and, 205–207
  verticals and, 184
Individual Capital Adequacy Standards (ICAS), 122–123
industrialised services, intangible services relationship with, 56, 60–61
information and communication technologies (ICTs), 91
  institutional specificity and, 77–78
  ISO standards and, 77
  service offshoring and, 62, 71, 77–78
information asymmetry, quality and, 63
information intensity, service offshoring and, 74
Information technology — IT Enabled Services/Business Process Outsourcing (ITESBPO) Lifecycle Processes. *See* ISO/IEC 30105
Infosys Technologies, 176
innovation hybrids, US and, 27–28
institution, quality as, 64
institutional continuum, 44, 46, 52, 110–111, 211
  data exchange standards and, 160
  hybrid actors spanning, 44–45
  insurance regulation and, 135–137
  life insurance and, 149–150
  Nasscom and, 199, 207–208
  service offshoring and, 57, 81–83
    India and, 177, 192–196, 205, 207
  sustainable reporting standards and, 169
institutional environments, 7–8, 86–91
  EU as, 86–87, 89–91, 94–99, 110–113
  ISO as, 87, 90–94, 110–113
  preferential trade agreements as, 103–107, 110–113
  US as, 86–87, 89–91, 99–103, 110–113
institutional specificity
  ICTs and, 77–78
  service offshoring hindered by, 72–73, 76–77
insurance, 13–14, 18, 21–22, 114–139, 212–213. *See also* European Directive Solvency II; European Insurance and Occupational Pensions Authority; International Association of Insurance Supervision; longevity risk; natural catastrophe insurance; Omnibus II Directive; reinsurance
  accumulation control and, 155
  AIG and, 122, 168–208
  basic capital requirements (BCR) and, 133
  consumer protection and, 137–138
  FIO and, 134
  governance and, 120–122
  institutional continuum and, 135–137
  market creation and, 22, 116–117, 140–172
  material continuum and, 135, 137–138

insurance (cont.)
micro-, 72
OECD countries and, 114–115
private insurers, self-regulatory power of, and, 44–45
regulation and, 116–139
risk-based regulation and, 123–125
service offshoring and, 72
sovereignty, as alternate form of, 14
spatial continuum and, 135, 138–139
spending on, 13
sustainable reporting standards and, 162–169, 171
US and, 114, 133–135, 138
insurance capital standard (ICS), IAIS and, 139
insurance core principles (ICPs), 133, 136
insurance premiums, 117–118
expenditure on, 114–116
life insurance and, 143
intangible services, 59
industrialised services relationship with, 56, 60–61
integrated production networks, 61
interest rates, life insurance and, 142–143
internal models, 126
capital requirements and, 136, 139
risk-based regulation and, 124
International Association of Insurance Supervision (IAIS), 24, 117, 137–139
BCR in, 133
ComFrame in, 133
G-SIIs in, 133
IAIGs in, 133
ICPs in, 133, 136
ICS and, 139
Solvency II and, 22, 132–135
International Business Machines Corporation. *See* IBM
International Financial Reporting Standards, 49–50
International Organisation of Securities Commissions (IOSCO), 50
International Organization for Standardization (ISO), 1–2, 21, 41, 61–62. *See also* ISO standards
COPOLCO and, 92
exclusion principle and, 93, 96
formal standards and, 51
hybrids and, 5
as institutional environment, 87, 90–94, 110–113
nanotechnology and, 47
service standards, distinctiveness of, and, 93–94
spatial continuum and, 84
standardization of services and, 88–94
internationalisation of services, 10–13. *See also* service offshoring
internationally active insurance groups (IAIGs), 133
Internet, qualification and, 65
INTERNORM, 18–19, 33–47
interoperability standards, 117, 140
IOSCO. *See* International Organisation of Securities Commissions
ISO. *See* International Organization for Standardization
ISO 9000, 103, 186, 198
ISO 14000, 8
ISO 45001, 12
ISO standards, 77, 92–94
ICTs and, 77
India, service offshoring in, and, 193–194
MSS in, 93–94
national participation and, 193
service related, percentage of, 92
ISO/IEC 20000, 203
ISO/IEC 30105 (Information technology — IT Enabled Services/Business Process Outsourcing (ITESBPO) Lifecycle Processes), 204–205
India, service offshoring in, and, 202–208, 220
IT Act, in India, 199
IT hardware industry, India, service offshoring in, and, 180
IT-enabled services (ITeS), 15, 173–174, 188–191, 196–198
ITESBPO. *See* ISO/IEC 30105
ITS. *See* Implementing Technical Standards

Japan
European Union Economic Partnership Agreement, 104–105
Jasanoff, Sheila, 219
Jegen, Maya, 213
JP Morgan, 122

Kalra, Rajesh, 183, 202
Kapur, D., 201
Karpik, Lucien, 58
Kenney, Martin, 177, 183
Kessler, Oliver, 216
Kilbreth, Jeff, 158
Kjaer, Poul, 26–27
Knafo, Samuel, 215
Knight, Franck H., 62–63
Koppell, Jonathan G., 27, 194–196

KPMG, 164
outsourcing and, 54, 174
Kraidy, Marwan, 40
Kshetri, N., 200
Kumar, Punit, 193

labour costs, service offshoring driven by, 10–11, 15, 52–55, 71, 173–174, 179
labour intensity of services, 17
language skills, service offshoring driven by, 71–72
Latour, Bruno, 37–38, 215–216
hybridity and, 6, 36–38
quasi-objects and, 6, 37–39, 47
law, private international authority and, 42
Lawsky, Benjamin, 134–135
Leander, Anna, 35
legitimacy, 4
Levi-Faur, David, 27
Lezon, Catherine, 138
liberalisation reforms, India, service offshoring in, and, 177, 181–182, 198
Life and Longevity Markets Association (LLMA), 148–149
life insurance
annuities and, 124, 144
institutional continuum and, 149–150
insurance premiums for, 143
interest rates and, 142–143
Lifemetrics and, 148–149
LLMA and, 148–149
longevity risk and, 141, 143–151, 170
longevity standard and, 146–151
market creation and, 140–170
material continuum and, 150
natural catastrophe insurance versus, 118
reinsurance and, 119
risk-based regulation and, 124, 142
securitisation and, 118–121, 142–151, 170
Solvency II and, 128–129
spatial continuum and, 150–151
Lifemetrics, 148–149
Lim, Alwyn, 165
Lipschutz, Ronnie, 5, 28
Lisbon Agenda, 95
LLMA. *See* Life and Longevity Markets Association
Lobo-Guerrero, Luis, 14, 121
long term guarantee measures, in Omnibus II, 127–132
longevity risk
life insurance and, 141, 143–151, 170
LLMA and, 148–149
OECD on, 149
UK and, 145–146
Xpect - Club Vita Indice and, 149
longevity standard, 146–151
institutional continuum and, 149–150
Lifemetrics and, 148–149
material continuum and, 150
spatial continuum and, 150–151
Lukes, Steven, 4

Macron, Emanuel, 213
Magnette, Paul, 105
Mallard, Grégoire, 213–214
Malmström, Cecilia, 103
management
BPM and, 175
corporate governance and, 30
service offshoring viewed via, 73–74
value-based, 30
management system standards (MSS)
China and, 195
India in, 186, 203, 205–206
ISO standards and, 77, 92–94, 186
mandatory standards, voluntary specifications versus, 7, 79
manufactured goods, service-content of, 69–70
market creation
GRI and, 141
insurance and, 22, 116–117, 140–172
life insurance and, 140–170
natural catastrophe insurance and, 141, 154–162
pension funds and, 140–151
uncertainty and, 63, 120–121
matching adjustments, in Solvency II, 128–130
material continuum, 83–84, 111–112, 211, 220–221
data exchange standards and, 160–161, 211
hybrid objects spanning, 46–47, 52
insurance regulation and, 135, 137–138
life insurance and, 150
Nasscom and, 199, 207
service offshoring and, 57
India and, 185–192, 205, 207
Mathis, James, 109
Mattli, Walter, 99–100
Mazzaferro, Francesco, 130
MCR. *See* minimum capital requirements
Means, Gardiner, 30
Mérand, Frédéric, 213
Mercier, Delphine, 79–80
MFN. *See* most-favoured-nation

micro-insurance, 72
Miles, Ian, 61
minimum capital requirements (MCR), 126
model providers, natural catastrophe insurance and, 158–159, 171
modernity, crisis of, hybridity and, 6, 36–37
Modi, Narendra, 174
Mohan, Rama, 197
Mondou, Matthieu, 219
Montgomerie, Johnna, 215
Morgan, Katharine E., 102
most-favoured-nation (MFN), GATS and, 108–109
Mouhoud, El Mouhoub, 76
MSS. *See* management system standards
multi-level governance, EU and, 32
Murphy, Craig, 9, 29
Munich Re, 155–158, 171
mutual recognition of standards, 85
  CETA and, 108
  EU and, 85
  preferential trade agreements and, 105–112
myths
  ambiguity, power of, and, 36
  hybrids and, 5–6, 35–36

NAFTA. *See* North American Free Trade Agreement
NAIC. *See* National Association of Insurance Commissioners
nanotechnology, ISO and, 47
Nasscom. *See* National Association of Software and Services Companies
National Association of Insurance Commissioners (NAIC), 133–134
National Association of Software and Services Companies (Nasscom), 23, 174, 176, 185, 198–202, 206
  BPM and, 175
  DSCI and, 200
  institutional continuum and, 199, 207–208
  ISO/IEC 30105 and, 202–208, 220
  IT Act and, 199
  land access and, 200
  material continuum and, 199, 207
  National Skill Registry and, 199–201
  spatial continuum and, 199
  US and, 201–202
National Institute of Standards and Technology (NIST), 100–102, 110
national participation
  EU and, 90, 110, 112
  ISO standards and, 193
National Skill Registry, in India, 199–201
National Technology Transfer and Advancement Act, 100
natural catastrophe insurance
  accumulation control and, 155
  Cresta and, 155–161
  data exchange standards and, 156–162, 171
  Fukushima Daiichi nuclear disaster and, 151
  life insurance versus, 118
  market creation and, 141, 154–162
  model providers and, 158–159, 171
  natural hazard models in, 155
  reinsurance and, 119, 141, 151–152, 154–162
natural hazard models, 155
New Approach, EU (Council Resolution 85/C 136/01), 21, 90, 94–95, 99
Ní Mhurchú, Aoileann, 39
Nilekani, Nandan, 176
NIST. *See* National Institute of Standards and Technology
non-tariff measures, 71–73
  service offshoring and, 72–73, 172
  Swiss Federal Act on Technical Barriers to Trade as, 81
  WTO on, 72–73, 81
North American Free Trade Agreement (NAFTA), 218
  Trump and, 104

OECD. *See* Organisation for Economic Co-operation and Development
offshoring. *See also* service offshoring
  outsourcing versus, 54
Omnibus II Directive (Directive 2014/51/EU), 125, 127–128, 138
  Delegated Regulation (EU) 2015/35 and, 125, 131–132
  long term guarantee measures in, 127–132
Organisation for Economic Co-operation and Development (OECD), 89
  longevity risk and, 149
  STRI by, 73
ORSA. *See* Own Risk and Solvency Assessment
orthography, of hybrids, 35
Osborne, George, 144
outsourcing, 15
  death of, 54
  KPMG on, 54, 174
  offshoring versus, 54
Own Risk and Solvency Assessment (ORSA), 126, 134, 136

Palan, Ronen, 50
Parliament of Things, 37–38, 47
Parthasarathy, Balaji, 177, 219
Pauly, Louis W., 45

Payne, Anthony, 44–45
pension buy-ins, 145
pension buy-outs, 145
pension funds, 13, 116
  buy-in arrangements in, 145
  buy-out arrangements in, 145
  market creation and, 140–151
  OECD countries and, 114–115
  securitisation and, 119–120, 142–151
  Solvency II and, 129, 137
  US and, 144
Petit, Pascal, 74–75
Phillips, Nicola, 44–45
Pijl, Kees van der, 40–41
platforms, service offshoring and, 3–4, 61–62, 159–160, 184–185, 221
post-colonial studies, 6, 35, 39–41, 53, 215
Potoski, Matthew, 8
power
  authority versus, 4
  governance as mode of exerting, 29
Prabhu, Anjali, 40
Prakash, Aseem, 8
preferential trade agreements. *See also* *specific* trade agreements
  civil society and, 106
  GATS and, 108–109
  as institutional environment, 103–107, 110–113
  mutual recognition of standards and, 105–112
  qualification and, 107–108, 112
  regulatory chill effect of, 105–106
  regulatory cooperation and, 106–107, 111, 211
  third countries and, 107–113
  WTO and, 108–109
principle-based regulation, 123, 134–135
private insurers, self-regulatory power of, 44–45
private international authority, 4–6, 42. *See also* transnational hybrid authority
  law and, 42
private/public distinction. *See* institutional continuum
process standards
  India, service offshoring in, and, 185–192
  ISO/IEC 30105 and, 202–208, 220

QIS5. *See* quantitative impact study, fifth, for Solvency II
Quack, Sigrid, 33–34
qualification, 85
  hybrid fora and, 65–66
  Internet and, 65
  professional, preferential trade agreements and, 107–108, 112
  quality versus, 64–66
quality, 62–66, 77
  EU and, 96
  India, service offshoring in, and, 182–191
  information assymetry and, 63
  as institution, 64
  ISO standards and, 186
  personalisation versus industrialisation based, 63
  qualification versus, 64–66
  standardisation, neo-Taylorist, and, 75
quantitative impact study, fifth, for Solvency II (QIS5), 127, 131
quasi-objects, 46
  Latour on, 6, 37–39, 47

Rao, Narasimha, 181
RCEP. *See* Regional Comprehensive Economic Partnership
regime-hybridity, 28
Regional Comprehensive Economic Partnership (RCEP), 103–104
regulation. *See also* risk-based regulation; self-regulation
  governance and, 19
  hybridity and, 26, 32
  insurance and, 13–14, 21–22, 116–139
  Lamfalussy process in, 125, 127
  standards and, 6–12, 24–26, 55–56, 80, 85
  standards as, 20–22, 91, 122, 135–136
Regulation on European Standardisation (1025/2012), 96, 99, 110
régulation theory, 9, 12, 64, 74–78
regulatory capture, 136
regulatory chill effect
  CETA and, 106
  of preferential trade agreements, 105–106
regulatory competition, 87
regulatory convergence, 3–4, 84, 86–87
  CETA and, 48, 104, 106–107
  WTO promoting, 89
regulatory cooperation
  CETA and, 106–107
  CPTPP and, 106
  mandatory versus voluntary, 107
  preferential trade agreements and, 106–107, 111, 211
regulatory state agencies, risk-based regulation and, 124
reinsurance, 117, 119, 134, 140, 170–171
  accumulation control and, 155
  Cresta and, 156–158, 171

reinsurance (cont.)
  life insurance and, 119
  Munich Re and, 155–158, 171
  natural catastrophe insurance and, 119, 141, 151–152, 154–162
  Swiss Re and, 146–147, 150, 154–158, 171
relational intensity, 17
  service offshoring and, 74
relational services. *See* intangible services
reputation, 45
reshoring, US and, 66, 71
restrictive hypothesis, 12, 211–212
  on service offshoring, 56–58, 76–78, 85, 173
risk society, 13, 36–37
risk-based regulation, 123–125
  as behavioural regulation, 123–124
  internal models and, 124
  life insurance and, 124, 142
  as principle-based regulation, 123
  private companies authority in, 124
  regulatory state agencies authority in, 124
  rule-based regulation versus, 123–124
  US and, 124
Roy, Raman, 182, 184, 193, 202
rule-based regulation, risk-based regulation versus, 123–124

Sahlin-Andersson, Kerstin, 45
Saïd, Edward, 39
Sassen, Saskia, 4
  hybrids and, 33
Schepel, Harm, 9, 79
science, technology and society studies (STS studies), 6, 35–39, 46, 64–66, 215–216
SCR. *See* solvency capital requirements
sectorial specificity. *See also* distinctiveness of service standards
  Afnor and, 97–98
  service offshoring and, 72, 76–78
securitisation, 119–121
  life insurance and, 119–120, 142–151, 170
  pension funds and, 119–120, 142–151
  SPEs and, 119
Segal, Rajesh, 197–198
SEI. *See* Software Engineering Institute
self-regulation, 136
  ambiguity of standards and, 217
  EU and, 95
  private insurers and, 44–45
semiotics, 5–6, 35–36
Senarclens, Pierre de, 31
service offshoring, 10–11, 20, 54–85, 173. *See also* India, service offshoring in; 75/25 puzzle
  cloud computing and, 62
  codification and, 74
  digitability and, 71
  education driving, 71–72, 181–182
  extensive hypothesis on, 56–58, 78–79, 85, 173
  foreign investment and, 55
  gender equality and, 75, 78
  ICTs and, 62, 71, 77–78
  information intensity and, 74
  institutional continuum and, 57, 81–83
  institutional specificity hindering, 72–73, 76–77
  insurance and, 72
  labour costs driving, 71
  language skills driving, 71–72
  managerial perspective on, 73–74
  material continuum and, 57, 83–84
  mobility of service providers driving, 71
  non-tariff measures and, 72–73
  platforms and, 3–4, 61–62, 159–160, 184–185, 221
  relational intensity and, 74
  restrictive hypothesis on, 56–58, 76–78, 85, 173
  sectorial specificity and, 72, 76–78
  social equality and, 75, 78
  social relation of accessibility and, 75–76, 78
  spatial continuum and, 57, 84–85
  standards-defying services and, 79–85
  transnational hybrid authority in, 56, 58
  US and, 55
services. *See also specific topics*
  difficulty defining, 58–60
  intangibility of, 59
  mode of request in, 60
  rise of, 3–4, 10–13, 88, 211–212, 219–220
  as transformation of the state of an individual or an object, 59–60
  United Nations Manual on Statistics of International Trade in Services defining, 59
Services Trade Restrictiveness Index (STRI), 73
75/25 puzzle, 20, 57–58, 211
  developing versus developed countries and, 66–67, 70
  domestic services and, 69–70
Shekar, Karthik, 201
Six Sigma method, 77, 191

smart phones, 61
social contestability of standards, 217
social equality, service offshoring and, 75, 78
social relation of accessibility, 161
  service offshoring and, 75–76, 78
socialisation of standards, 218
Society for Worldwide Interbank Financial Telecommunication (SWIFT), 153
Software Engineering Institute (SEI), 187
Software Technology Parks (STPs), India, service offshoring in, and, 181–182, 200
solvency capital requirements (SCR), 124–128
Solvency II. *See* European Directive Solvency II
Solvency II Framework Directive, 125
sovereignty
  dual nature of, 50
  governance and, 29
  insurance as alternate form of, 14
  insurantial, 121
  The Stack as form of, 221
spatial continuum, 112–113, 211
  data exchange standards and, 161
  hybrid spaces spanning the recognition of standards, 49–52
  insurance regulation and, 135, 138–139
  ISO and, 84
  life insurance and, 150–151
  Nasscom and, 199
  service offshoring and, 57, 84–85
    India and, 196–198, 205–208
special purpose vehicles (SPVs), 119
Spence, Michael, 63
SPVs. *See* special purpose vehicles
Srinivas, Smita, 177
Srinivasan, Janaki, 219
standardisation, neo-Taylorist, 75, 220
standards. *See* formal standards; *specific topics*
standards-defying services, service offshoring and, 79–85
state
  internationalisation of the, 9
  recognition, transnational hybrid authority and, 5, 42, 48–51, 209–210
state recognition, transnational hybrid authority and, 45–46
Stiglitz, Joseph, 63
STPs. *See* Software Technology Parks
Strange, Susan, 14, 42, 120, 214
STRI. *See* Services Trade Restrictiveness Index
substantive attribute
  ambiguity as, 19, 25, 34–35, 52, 210
  hybridity as, 25, 34, 210
Sunstein, Cass, 123–124
sustainable reporting standards
  Bermuda and, 168–169
  institutional continuum and, 169
  insurance and, 162–169, 171
  US and, 169
SWIFT. *See* Society for Worldwide Interbank Financial Telecommunication
Swiss Federal Act on Technical Barriers to Trade, 81
Swiss Re, 146–147, 150, 154–158, 171

Tamm Hallström, Kristina, 186
Tata Consultancy Services (TCS), 179
Taylor, Dough, 137
TBT Agreement. *See* Technical Barriers to Trade Agreement
TCS. *See* Tata Consultancy Services
technical authority, 47
Technical Barriers to Trade Agreement, WTO (TBT Agreement), 88–89, 218
technological change, transnational hybrid authority and, 46–47
technological innovation, standards impacting, 45–46, 79–80
  Indian IT services in, 178–182
territorial sovereignty. *See also* spatial continuum
  hybridity and, 41
  standards replicating, 50–51, 112
tertiarisation, test of, 58–66. *See also* services, rise of
third countries, in preferential trade agreements, 107–113
Timmermans, Stefan, 9
trade unions, India, service offshoring in, and, 200–201, 207
Transatlantic Trade and Investment Partnership (TTIP), 3–4, 103–105
  Trump and, 104
transnational hybrid authority, 2–3, 7–10, 12, 43, 80–82, 210. *See also* institutional continuum; material continuum; spatial continuum
  consent and, 45
  hegemony and, 216
  hybrid actors and, 43–46
  hybrid objects and, 46–48
  hybrid spaces and, 48–51
  hybrids and, 53
  service offshoring and, 56, 58

transnational hybrid authority (cont.)
India and, 206–207
state recognition and, 5, 42, 45–46, 48–51, 209–210
technological change and, 46–47
transnational private authority, 4–6, 43–46
Treaty on Stability, Coordination and Governance, 32
tripartite standards regime, 81
Trump, Donald, 104
Tsutsui, Kiyoteru, 165
TTIP. *See* Transatlantic Trade and Investment Partnership

UN/CEFACT. *See* United Nations Centre for Trade Facilitation and Electronic Business
uncertainty, market creation and, 63, 120–121
UN/EDIFACT. *See* United Nations Electronic Data Interchange for Administration, Commerce and Transport
UNICEDE. *See* Universal Cession Electronic Data Exchange
United Kingdom (UK), 129
annuities and, 144
Brexit and, 103–104
longevity risk and, 145–146
United Nations Centre for Trade Facilitation and Electronic Business (UN/CEFACT), 153–154
United Nations Electronic Data Interchange for Administration, Commerce and Transport (UN/EDIFACT), 153
United Nations Manual on Statistics of International Trade in Services, 59
United Nations Trade Data Interchange Directory (UNTDID), 153
United States (US), 21. *See also* American National Standards Institute
ASME and, 101–103
ASTM International and, 102–103
capital requirements and, 135
civil society in, 101
corporate social responsibility and, 163–164
data exchange standards and, 161
direct participation in standardisation and, 89, 110, 112, 193–194
energy standards and, 111–112
FIO in, 134
innovation hybrids in, 27–28
as institutional environment, 86–87, 89–91, 99–103, 110–113
insurance and, 114, 133–135, 138
international standards recognized in, 101
Nasscom and, 201–202
National Technology Transfer and Advancement Act and, 100
NIST and, 100–102, 110
pension funds and, 144
principle-based regulation and, 134–135
reshoring and, 66, 71
risk-based regulation and, 124
service offshoring and, 55
sustainable reporting standards and, 169
US GAAP and, 135, 139
United States Standards Strategy (USSS), 101
Universal Cession Electronic Data Exchange (UNICEDE), 158
UNTDID. *See* United Nations Trade Data Interchange Directory
US. *See* United States
US Generally Accepted Accounting Principles (US GAAP), 135, 139
USSS. *See* United States Standards Strategy

VanDuzer, J. Anthony, 109
Veblen, Thorstein, 49
venture capitalists, 74–75
vertical service standards, 98–99, 111, 212
verticals, India, service offshoring in, and, 184
Villar, Costanze, 213
Vion, Antoine, 79–80
Vogel, David, 107
volatility adjustments, in Solvency II, 122–131
voluntary specifications, mandatory standards versus, 7, 79

W3C. *See* World Wide Web Consortium
Weiss, Linda, 27, 194–195
Winickoff, David E., 219
World Trade Organization (WTO), 88–89
non-tariff measures and, 72–73, 81
preferential trade agreements and, 108–109
regulatory convergence promoted by, 89
TBT Agreement by, 218
World Wide Web Consortium (W3C), 152–153
WTO. *See* World Trade Organization

XML. *See* Extensible Markup Language
Xpect - Club Vita Indice, 149

Yates, JoAnne, 9
Young, Robert, 39

Zhang, Vicki, 14, 118–119, 123